Microchips

The Illustrated Hitchhiker's Guide to Analytical Microchips

Larry J. Kricka, DPhil, FACB, FRSC CChem, FRCPath
Department of Pathology and Laboratory Medicine
University of Pennsylvania Medical Center
Philadelphia, Pennsylvania

2101 L Street, NW, Suite 202
Washington , DC 20037-1558

1 2 3 4 5 6 7 8 9 0   GG   04 03 02

Printed in the United States of America

Library of Congress Cataloging-in-Publication Data

Kricka, Larry J., 1947–
    Microchips : the illustrated hitchhiker's guide to analytical microchips / Larry J. Kricka.
      p. cm.
    Includes index
    ISBN 1-890883-75-1 (alk. paper)
      1. Biochips.   I. Title.

R857.B5 K753 2002
610'.28–dc21

                                                                      2002066714

Microchips

To Barbara
for her patience and understanding
these many years!

Preface

Analytical microchips in their different guises as microfluidic chips, biochips, bioelectronic chips, and microarray chips are having a profound impact on the way in which analyses are performed in laboratories worldwide. These ultra–small analyzers are constructed using microfabrication techniques, such as photolithography and ink-jetting, adapted from the microelectronics and printing industry. They offer simple and intriguing new ways of performing analyses in which the successive steps in a conventional analytical process are integrated onto small "chips" of silicon, glass, or plastic to produce a "laboratory on a chip" or a "miniaturized total chemical analysis system" (μ-TAS). The new analytical chips require only microliter or sub-microliter volumes of sample and reagents, and promise substantial cost reductions and greatly improved ease of use.

Many of the conventional analytical techniques have been miniaturized to a microchip format including capillary electrophoresis (CE), liquid chromatography (LC), immunoassay, polymerase chain reaction (PCR) amplification reactions, and flow cytometry. A prospect for the future is the design and manufacture of hybrid devices that provide total integration with all components of the analyzer (electronic control circuits, sample processing, analytical structures, reagents, communications) on a single chip-like device. Already, the microchip toolbox contains a vast collection of different components, such as microchannels, microchambers, microelectronics, electric motors, pumps, valves, refrigeration, heaters, lasers, optics, and detectors. In theory, all these components are available for use in the construction of increasingly sophisticated analytical microchips.

More frequent encounters with these new inhabitants of our laboratory world will no doubt prompt questions or spark the inquisitive to seek out more information. This guide was prepared with these eventualities in mind. It contains information on the key types of microchips and the processes important in their fabrication and operation, and other more lighthearted aspects of the microchip world. It also includes glimpses of the next step in miniaturization, namely nanotechnology. Devices with nm-sized components are confidently expected to provide the next generation of analyzers. Progress in the analytical applications of nm-diameter carbon nanotubes is just one tangible example of this future trend.

Recommended starting points in this guide are "M" for "microchip" and "M" for "microminiaturization." Here you will find a general introduction that will serve as a pointer to other entries in the guide. Microchips have been the victim of uncontrolled and non-systematic nomenclature, and this has caused its share of confusion. Originally, the term "microchip" referred to a microelectronic device, but now it also refers to the different types of analytical microchips. Efforts are currently underway to provide a systematic basis for naming the different types of analytical microchips, and this should alleviate the confusion in the literature.

I hope that this guide will be a useful introduction to the general microchip area, and will prompt you to explore the broader aspects of microchips, and the myriad of applications of these Lilliputian devices.

Acknowledgments

I thank my colleagues in the microchip world for supplying photographs of their microchips, and Kevin Dudek, Dean Cummings, and Paul Schiffenmacher at Biomedical Communications at the University of Pennsylvania for preparing many of the illustrations and providing the front cover design. I also wish to thank Peter Wilding, David Goodman, Philip Stanley, and Paolo Fortina for their thoughtful and helpful review of the manuscript of this book.

Front cover design "Microchips on the head of a pin" (Kevin Dudek and Dean Cummings) features a selection of PCR microchips designed by Peter Wilding and Larry J. Kricka in the Department of Pathology and Laboratory Medicine, University of Pennsylvania Medical Center, Philadelphia, PA.

AACC

The American Association for Clinical Chemistry provides national and worldwide leadership in advancing the practice and profession of clinical laboratory science and its application to health care (see http://www.aacc.org). Please visit the AACC's Lab Tests Online website at www.labtestsonline.com if you are interested in learning more about the clinical aspects of the testing described in this guide. The long-running AACC Oak Ridge and San Diego Conferences have showcased many aspects of microchips and nanotechnology—relevant conference proceedings can be found in the journal *Clinical Chemistry* (1993; 39:705–42 and 1994; 40:1797–861).

Ångstrom

A unit of length named after the Swedish physicist Anders Jonas Ångström (1814–1874). One angstrom is equal to 0.1 nm (10^{-10} m). Dimensions of some common objects are compared in Figure 1, and for a fascinating pictorial journey through the universe of objects ranging from the vast to the smallest scale yet explored, see the excellent book entitled *Powers of Ten* (*1*).

Anodic bonding

This process, also known as electrostatic or field-assisted bonding, is used to attach a piece of polished Pyrex glass to a polished silicon chip or an entire silicon wafer. It is important that the two surfaces to be bonded are completely flat. The silicon is placed onto a metal electrode (anode), and a Pyrex glass cover is positioned on top of the silicon (Figure 2). Electrostatic attraction pulls the two surfaces into close contact. The entire assembly is heated to 180–500 °C, and the Pyrex glass is contacted with the cathode at an applied direct current voltage of 200–1000 V. A strong electrostatic field develops between the two surfaces, and it pulls the surfaces into intimate contact. During the process, oxygen is transported out of the glass to the glass–silicon interface and reacts with the silicon to form a chemical bond between the two surfaces. It is important that the thermal-expansion coefficients of the two materials to be bonded are similar in order to minimize residual stress after bonding. The transparent cover of a glass-capped silicon microchip provides a simple means of observing events occurring within minute chambers and channels etched into the surface of the silicon chip.

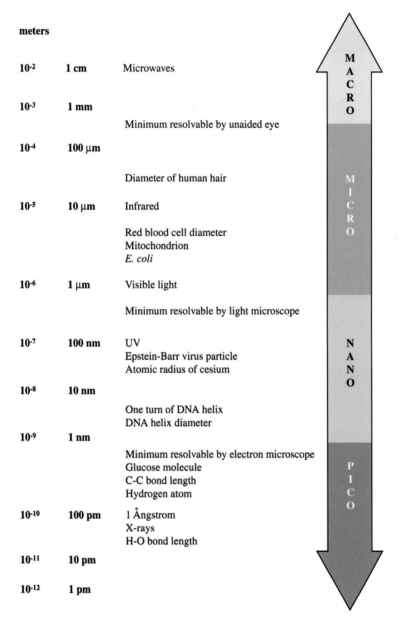

meters

10^{-2}	1 cm	Microwaves
10^{-3}	1 mm	
		Minimum resolvable by unaided eye
10^{-4}	100 µm	
		Diameter of human hair
10^{-5}	10 µm	Infrared
		Red blood cell diameter Mitochondrion *E. coli*
10^{-6}	1 µm	Visible light
		Minimum resolvable by light microscope
10^{-7}	100 nm	UV Epstein-Barr virus particle Atomic radius of cesium
10^{-8}	10 nm	
		One turn of DNA helix DNA helix diameter
10^{-9}	1 nm	
		Minimum resolvable by electron microscope Glucose molecule C-C bond length Hydrogen atom
10^{-10}	100 pm	1 Ångstrom X-rays H-O bond length
10^{-11}	10 pm	
10^{-12}	1 pm	

MACRO

MICRO

NANO

PICO

Figure 1. Size comparisons—from centimeter to picometer and beyond.

Array [See "In situ synthesis"; "Microarray"; "Tissue microarrays"]

In biomedical sciences it is often advantageous to carry out many tests simultaneously. Multi-analyte assays can be conveniently performed by arraying the reagents (or the test substances) as small discrete areas ("spots") on a flat surface of a microscope slide or a small piece of plastic or silicon (Figure 3). An array is usually produced by spotting or

2

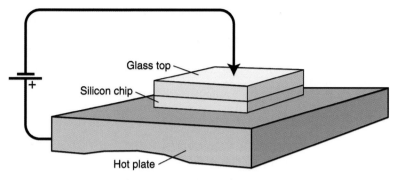

Figure 2. Experimental set-up for anodic bonding a Pyrex glass top to a silicon chip.

pipetting individual reagents directly onto the surface of a substrate (Figure 3A). Alternatively, an array of oligomeric reagents (oligonucleotides, peptides) can be built monomer-by-monomer on the substrate surface using in situ synthesis procedures. This array is then exposed to the test solution, and locations on the array are scanned to determine if a reaction with a component of the test solution has occurred (Figure 3B). The arrayed substances can be oligonucleotides, cDNA, antibodies, peptides, cell clones, or even small pieces of tissue. The density of the arrayed substances varies widely, and densities as high as 1 million/cm^2 can be achieved using in situ synthetic procedures.

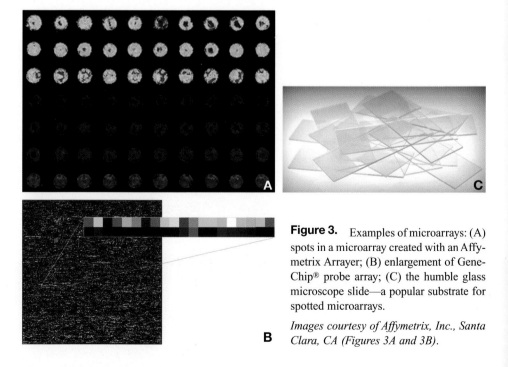

Figure 3. Examples of microarrays: (A) spots in a microarray created with an Affymetrix Arrayer; (B) enlargement of Gene-Chip® probe array; (C) the humble glass microscope slide—a popular substrate for spotted microarrays.

Images courtesy of Affymetrix, Inc., Santa Clara, CA (Figures 3A and 3B).

Arrayer [See "Spotter"]

A device for placing small drops of solutions containing probes or targets onto the surface of a substrate to form a microarray. A pipette or ink-jet device is usually used to deliver the drop onto the surface.

Aspect ratio [See "LIGA"]

Ratio of the height to the width of a microfabricated feature such as a microchannel.

Atomic force microscope [See "Microcantilever"]

The atomic force microscope (AFM) has revolutionized microscopy and revealed the world at the atomic and molecular level. The heart of an AFM microscope is a small sharp tip at the end of a silicon-micromachined microcantilever. In contact type of AFM, the μm–nm-sized cantilever tip is dragged across a surface. The vertical motion of the tip as it traverses the surface is detected and converted into a three-dimensional map of the surface. If the tip is coated with molecular groups, then the AFM can detect specific interactions between the tip and molecules on the surface of the sample.

Attomole

An attomole is 10^{-18} moles. Microminiaturization has hurled us headlong into realms inhabited by very few molecules. One attomole is about 600,000 molecules (Avogadro's number/10^{-18}). Below this amount lies the zeptomole (10^{-21}), about 600 molecules, and eventually the yoctomole (10^{-24}). This is just a fraction of a molecule—now you see it, now you don't (just ask Heisenberg)!

Bead array

An array of reagent-coated microbeads at the tip of a fiber-optic bundle provides a powerful tool for simultaneous high-throughput analysis. The array is constructed by first etching the end of the bundle to produce a well at the end of each fiber, large enough to contain a polystyrene microbead (3 μm in diameter) (Figure 4A). Next, a pool of beads is prepared from different batches of beads. The beads in each batch are treated with a mixture of fluorescent dyes to give the beads a unique fluorescence signal, and are then coated with a different reagent (e.g., an oligonucleotide probe). The bead array is constructed by drawing the beads up into the wells at the end of the fibers, one bead per well. Each bundle contains tens of thousands of fibers (e.g., a 1-mm diameter bundle has 50,000 individual fibers) (Figure 4B), and individual bundles can be assembled into cassettes with conventional 96-, 384-, or 1536-well microplate geometries (Figures 4C and 4D). The bundles are dipped into the test solution(s) to allow target molecules in the test sample to hybridize to probes bound to the microbeads. Fluorescence excitation light is piped to the bead held in the well at the end of the fiber. The fiber collects fluorescence emission signal from the bead so that the identity of the bead can be decoded from the dye signature. Fluorescent signal is also collected from any fluorophore-labeled target that has bound to the probe on the bead surface. Ultimately, this bead array strategy will allow millions of tests to be performed simultaneously in a single experiment.

Biochip

This term was originally applied to biological versions of microelectronic chips (2), the notion being that biological molecules could be assembled into a device with the same computational capabilities as a silicon-based microelectronic chip. Subsequently, it has been applied in a more general way to microchips used to analyze biological fluids or manipulate the components of a biological fluid.

Bioelectronic microchip [See "Electronic hybridization"; "Electronic stringency"]

This is a microchip designed for the analysis of biological fluids. It incorporates microelectronic components, e.g., electrodes, which can manipulate components of the biological sample contained in chambers or channels within the chip. Figure 5 shows some examples of bioelectronic chips that contain arrays of μm-sized electrodes. These elec-

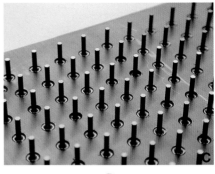

Figure 4. Bead microarrays: (A) microbeads (3 μm in diameter) on surface of fiber-optic bundle; (B) close-up of fiber-optic bundle; (C) array of fiber optics; (D) cartridge containing the array of fiber optics in a 96-well format.

Reproduced with permission from Illumina, Inc., San Diego, CA.

trodes are used to manipulate cells or other molecules contained in the sample introduced into the microchip.

Bioinformatics

A part of information science that is dedicated to the collection, manipulation, and analysis of biological data. The advent of massively parallel assay formats (e.g., microarrays) has led to individual experiments that produce enormous amounts of data (e.g., meas-

---

Figure 5. Examples of bioelectronic chips: (A) i-STAT blood analysis cartridge (sodium, potassium, glucose, ionized calcium, pH, hematocrit, hemoglobin); (B) top (left) and underside of a NanoChip™ (Nanogen, Inc., San Diego, CA); (C) close-up of electrical connections for the array of microelectrodes in a NanoChip™; (D) microtransponder (an oligonucleotide probe-coated integrated circuit composed of photocells, memory, clock and antenna); (E) AVIVA cell manipulation biochip; (F, G) old and new generation of the Motorola Life Sciences eSensor™ DNA biochips; (H) eSensor™ biochip opened for sample addition.

Reproduced with permission from i-STAT Corporation, East Windsor, NJ (Figure 5A); PharmaSeq, Inc., Monmouth Junction, NJ (Figure 5D); AVIVA Biosciences Corporation, San Diego, CA (Figure 5E); Motorola Life Sciences, Pasadena, CA (Figures 5G and 5H).

A

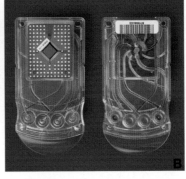

B

C

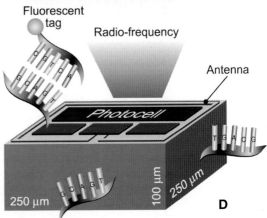

Fluorescent tag

Radio-frequency

Antenna

Photocell

250 μm

100 μm

250 μm

D

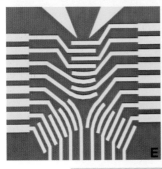

E

F

G

H

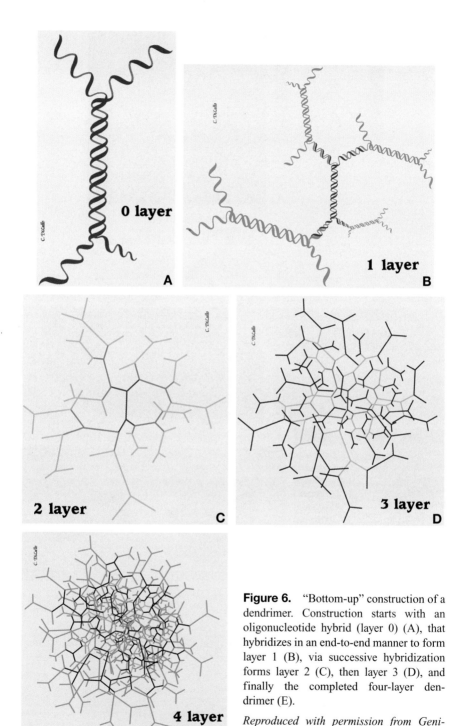

Figure 6. "Bottom-up" construction of a dendrimer. Construction starts with an oligonucleotide hybrid (layer 0) (A), that hybridizes in an end-to-end manner to form layer 1 (B), via successive hybridization forms layer 2 (C), then layer 3 (D), and finally the completed four-layer dendrimer (E).

Reproduced with permission from Genisphere, Bala Cynwyd, PA.

urements are made at two wavelengths at each of tens of thousands of spots in a two-color microarray experiment). Not surprisingly, this has presented data analysis tasks and interpretation on a scale not encountered previously in the biological sciences.

Bonding [See "Anodic bonding"]

This is the term used to describe the process of joining two chip components together, e.g., an etched silicon or glass microchip and a glass cover. Many types of bonding processes have been developed to bond microchip components such as anodic bonding (glass to silicon component), fusion bonding (silicon to silicon, or silicon to silicon via an intermediate thin layer of glass), and eutectic bonding (silicon to silicon via an intermediate thin layer of gold).

"Bottom-up" construction [See "Carbon nanotube"; "Nanotechnology"]

A process that builds a device one atom or molecule at a time is referred to as "bottom-up" construction. This contrasts with the "top-down" approach, in which a block of material is fabricated into the desired device by removing material through etching, ablation or a machining process. Typically, microdevices are made by the top-down process, whereas nanodevices are made by the bottom-up construction process. A good example of a "bottom-up" device is a multi-component oligonucleotide dendrimer. This is made layer-by-layer from oligonucleotide building blocks that are designed to hybridize end-to-end with each other, as shown in Figure 6. The hybridization reactions form successive layers of the large and open dendrimeric structure. This nanotechnological object provides a glimpse of the possibilities that lie ahead for nanochips that could be assembled atom-by-atom or molecule-by-molecule to produce functional and useful analytical devices.

Capillary electrophoresis (CE)

This is the type of electrophoresis that has been most commonly adapted to a microchip format. The 25- to 100-cm long, flexible, 25- to 100-μm diameter silica capillaries used in conventional CE are replaced with much shorter microchannels etched in a small flat piece of glass or plastic. The layout of a multiple-sample CE chip is shown in Figures 7A and 7B. It comprises a glass chip with a series of microchannels (typically 50-μm wide × 10-μm deep). The chip is sealed with a plastic top to create microfluidic conduits, the ends of which connect to open reservoirs for introduction of fluids. The microchip is operated by a tabletop-sized analyzer (Figure 7C and 7D). The analytical output from the chip is either an electropherogram of the separation (Figure 7E) or a gel-like image (Figure 7F). The current family of microchips for DNA analysis can size DNA in the range 25–12,000 bp using a 1-μL sample size (12 assays per chip, sieving polymer-filled microchannel, laser-induced fluorescence detection).

The advantages of performing CE in a microchip include valveless injection, zero dead volumes, and efficient dissipation of heat, which permits the use of higher field strengths and hence faster separations.

Cartridge

The small physical size of a microchip, typically less than 1 cm × 1 cm and a few millimeters thick, makes handling them awkward and difficult. Embedding a microchip in a larger cartridge or a cassette is a convenient way of making it easily handled and manipulated. It also provides a framework to mount connectors that provide fluidic access into the interior of the microchip.

cDNA

Complementary DNA (cDNA) is produced from an RNA template using the enzyme reverse transcriptase (the enzymatically synthesized cDNA is complementary to the RNA). cDNAs representing different genes are used to make microarrays for the comparative assessment of gene expression.

cDNA array [See "Expression monitoring"]

cDNA molecules can be deposited onto the surface of a microscope slide in an ordered pattern to form a cDNA microarray. Each cDNA represents a gene, and this type of array is used to study gene expression in cells.

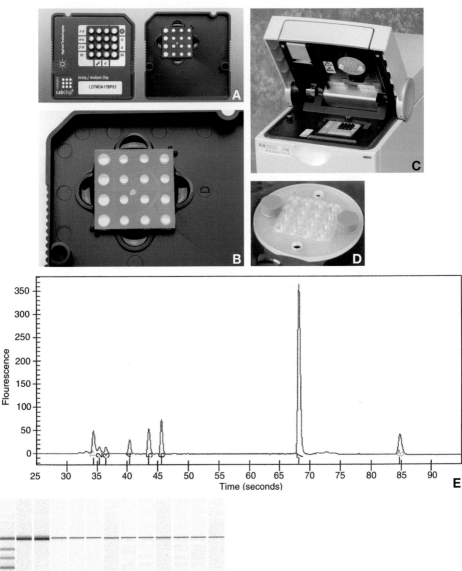

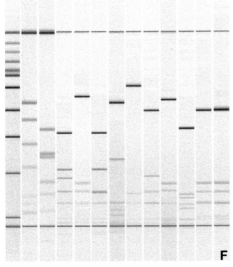

Figure 7. Microchip capillary electrophoresis: (A) top and underside of a LabChip® (Caliper Technologies Corporation); (B) close-up of the capillary electrophoresis microchannels; (C) Agilent 2100 Bioanalyzer: the analyzer is open to show the array of electrodes (underside of lid) that contacts with the LabChip®; (D) close-up of an electrode array that contacts with a LabChip®; (E) typical results of a DNA sizing experiment in the form of an electropherogram; (F) a gel-like image.

Cell isolation [See "Dielectrophoresis"]

Microchips and bioelectronic chips offer a number of highly effective ways of separating and isolating cells. Simple arrangements of posts are effective for filtering cells in microchannels—large cells cannot pass between the posts and they get trapped at or between the posts (Figure 8). Alternatively, a silicon weir structure (Figure 9) isolates cells by entrapment in the μm-wide gap between the top of the weir and the underside of the glass top capping the chip. Even greater control over cells can be exerted using electrical fields created by electrodes and arrays of electrodes within microchannels (Figure 10). Microelectrode-generated electrical fields in microchips are effective for trapping cells, isolating and selecting cells, and also lysing cells.

Chemiluminescence

Light emission produced in a chemical reaction is known as chemiluminescence. Chemiluminescent reactions are very sensitive and are used to detect targets hybridized to probes on microarrays. Common chemiluminescent detection methods include the luminol-based assays for horseradish peroxidase and adamantly 1,2-dioxetane reactions for alkaline phosphatase-labeled targets.

Chip

A term originally used to describe a minute square of thin semiconducting material (e.g., silicon) on which an integrated circuit is laid out. In the popular press and elsewhere, its usage has now been extended to include the different types of analytical microchips. If you are searching for information on analytical microchips and use "chip" as a search term, you will soon become familiar with "channel-forming integral protein" or CHIP for short. This is the archetypal member of the Aquaporin family of water channels, and probably not what you wanted!

Chipeener

Not a swashbuckling microchip scientist; instead a high-heeled shoe.

Chiplet

A little chip!

Chipmunk

Nothing to do with microchips; instead our furry friend the ground-squirrel (*Tamias striatus*).

Chipper

Not someone who makes microchips; instead it means lively, cheerful, "chirpy," or to twitter, chirp, babble, or chatter.

Figure 8. Silicon micropost-type filters and filter designs: (A) offset array of 20-μm posts spaced 7 μm apart in a 500-μm wide × 20-μm deep microchannel; (B) array of 73-μm wide microposts separated by 7-μm wide tortuous channels in a 500-μm wide × 5.7-μm deep microchannel; (C) filter microchip with three designs of flow deflectors and serial filters; (D) isolation of 5.78-μm diameter latex beads by a post-type filter (5-μm channels between 73-μm wide posts set across a 500-μm wide × 5.7-μm deep channel); (E) comb-type filter formed by an array of 120 posts (175-μm long × 18-μm wide) separated by 6-μm wide channels set across a 3-mm wide × 13-μm deep silicon channel; (F) stained white blood cells isolated by a comb-type filter (shows cells released from front of filter by reversing the flow through the filter).

Reproduced with permission from Anal Biochem *1998;257:95–100 (Academic Press/Elsevier).*

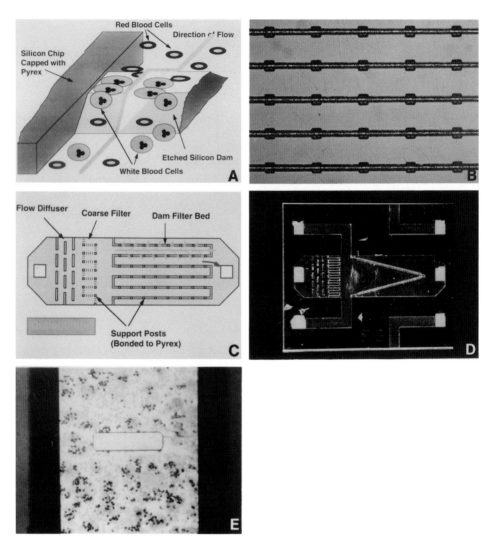

Figure 9. Weir-type silicon microfilters: (A) schematic representation of weir-type filter (3.5-μm gap between top of the etched silicon weir and underside of Pyrex glass top provides size-based filtration of cells); (B) silicon weir (nine 7.2-mm long weirs separated by 22-μm × 40-μm support posts); (C) filter microchip incorporating a coil-shaped filter bed; (D) filter microchip incorporating a V-shaped filter bed; (E) stained white blood cells filtered by a weir-type filter. The cells are trapped on top of the filter beneath the underside of the Pyrex glass cover on the microchip.

Reproduced with permission from Anal Biochem *1998;257:95–100 (Academic Press/Elsevier).*

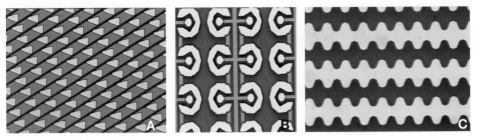

Figure 10. Bioelectronic microchips for cell isolation: (A) triangular electrode design; (B) point-circle electrode design; (C) sinusoidal electrode design.

Reproduced with permission from AVIVA Biosciences Corporation, San Diego, CA.

Chippies

Not casually dressed, free-living, flower-bedecked lovers of microchips; instead carpenters in the British naval air service.

Chippiness

Not a title for a regal microchip mogul; instead it means a state of being chippy, or short-tempered.

Chippy

This is not a description of a microchip in a euphoric state; instead it is the name of a shop that sells "fish and chips." Or the unpleasant physical sensations experienced after excessive consumption of alcoholic beverages.

Chips

Plural form of chip, but better known as the most popular type of fried potatoes, as in the British favo(u)rite "fish and chips."

Cliches

Science writers quickly seized the opportunity to coin amusing headlines that exploit the size of a microchip, the alternative meanings of the word "chip," or in the case of analytical microchips, their impact on the analytical sciences. Some examples are:

"Chips off the old block" (Reuter's News Report);

"Chip off ol' computer" (Philadelphia Daily News);

"Chip off the old chip" (Philadelphia Inquirer);

"The chipping forecast" (Nature Genetics);

"Microchip arrays put DNA on the spot" (Science);

"Biochips and DNA chips—The camel's nose is under the tent" (The Genesis Report);

"Let the chips fall where they may" (www.internetnews.com);

"Test tube's end" (Journal of Biomolecular Screening).

Colossus [See "ENIAC"; "Microcomputers"]

Along with ENIAC, this was one of the first computers to be built. Design and construction were completed in 1943, and the first of several colossi were installed at Bletchley Park in the United Kingdom (UK) in 1944. They were used very successfully to decipher coded messages during World War II (see http://www.codesandciphers.org.uk/lorenz/colossus/htm).

Confocal laser scanning microscopy

Detecting and recording an analytical signal from the μm-sized discrete areas on a microarray require specialized equipment and techniques. The resolution and sensitivity of confocal laser scanning microscopy is ideally suited for scanning microarrays and measuring fluorescence signal from individual spots on the array. Detection limits as low as 25 fluorescein-labeled molecules/mm^2 in a 50-μm^2 scanning area are attainable.

Controversy

Few areas of life are free from conflict or dispute, and microchip technology is no exception. Extensive patenting has led to costly litigation over ownership and infringement of intellectual property. The extent and complexity of litigation in the microarray field led one author to construct a helpful map to illustrate the inter-relationships between different commercial and academic adversaries (3).

Conventions [See "Probe"; "Target"]

Critters on a chip

A light-sensitive, 2-mm $\times$ 2-mm silicon chip coated with a genetically engineered bioluminescent bacteria (e.g., *Pseudomonas fluorescens HK44*) can detect the presence of specific substances, such as the common pollutant, naphthalene. These hybrid devices are sometimes referred to as "bioluminescent bioreporter integrated circuits" and produce light emission when the target substance is metabolized by the bacteria on the microchip.

Cyanine dyes

The cyanine dyes are widely used as labels in two-color assays, especially on cDNA-based microarrays. Usually, the orange fluorescing Cy3 [excitation maximum (ex) 550 nm, emission maximum (em) 570 nm] and the far-red fluorescing Cy5 (ex 647 nm, em 670 nm) are used in tandem.

Databases

Microchip literature has expanded rapidly, but luckily there are now several web-based sources and compilations of the microchip literature in the form of databases. These include www.gene-chips.com (bibliography and general information on all aspects of microarrays):

http://arrayit.com/e-library (microarray literature);

http://www.clinchem.org/cgi/content/full/47/8/1479/DC1 (microarray literature);

http://www.clinchem.org/cgi/content/full/48/4/662/DC1 (nanotechnology literature);

http://www.affymetrix.com/community/publications/index.affx (scientific papers and reviews published by Affymetrix);

http://www.bsi.vt.edu/ralscher/gridit/bioinformatics.htm (links and resources for microarray technology).

Dendrimer [See "Bottom-up construction"; "Nanotechnology"]

A dendrimer is a sphere-shaped or globular polymer molecule having a regular branched structure. Dendrimers are synthesized using branched monomers that react successively to produce layer upon layer (generations) of a tree-like structure. Their high surface functionality has made them particularly useful in DNA probe assays and immunological detection reactions. This is because the dendrimer serves as a site for the attachment of many labels, and this increases overall assay sensitivity.

Dielectrophoresis

The behavior of cells in a non-uniform alternating current (AC) field is governed by the dielectric properties of the individual cells. This parameter differs between different types of cells and the same type of cells in different physiological states. Dielectrophoretic forces can be thus used to trap, separate, and levitate cells according to the cell's intrinsic dielectric properties. Examples of cell separations achieved by dielectrophoresis in chips containing microelectrodes include separating viable from non-viable cells, and separating bacteria from blood cells.

Digital light processing [See "Light-directed synthesis"]

Micromirror arrays consist of over 1 million 16-µm × 16-µm microfabricated mirrors on a 4-cm × 4-cm chip. Each mirror can be tilted individually by 10° and thus it becomes possible to control the direction of light reflected from all the mirrors in the array simultaneously (Figures 11A–11C). These chips are used in consumer products for projection televisions, but have been adapted to make microarray chips. In this application, the micromirrors project light beams onto very small and specifically addressable sites on a microchip surface in order to initiate a photochemical synthesis reaction at those sites—the so-called digital light-processing process (Figure 11D). Successive rounds of photochemical reactions using a series of monomers, such as the four nucleotide bases (A, T, G, C),

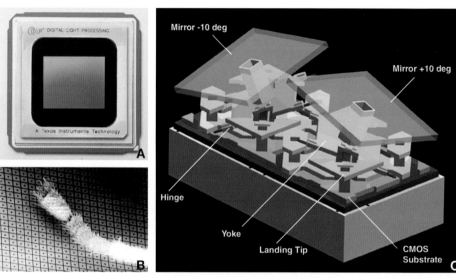

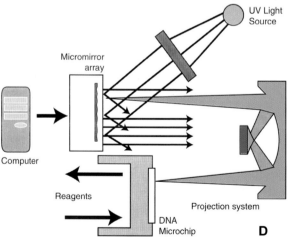

Figure 11. Digital micromirror array for digital light processing (DLP™): (A) DLP™ chip (SXGA DMD™, Texas Instruments) consisting of 1,310,720 individual 16-µm × 16-µm micromachined micromirrors; (B) Ant leg on the surface of the micromirror array; (C) components of the micromirror; (D) application of a digital micromirror array in light-directed synthesis to make oligonucleotide microarray chips (light is directed by individual mirrors to precise locations on the microchip surface to initiate oligonucleotide synthesis).

Reproduced with permission from Texas Instruments Inc., Dallas, TX (Figures 11A–11C).

builds up a series of oligonucleotides at the different locations to form an oligonucleotide microarray.

DNA

Deoxyribonucleic acid (DNA), as the storage site for the blueprint of our very being, makes this arguably the most important molecule. Please see "DNA Simplified II" for an informative and amusing guide to "everything you always wanted to know about DNA" (*4*).

DNA array [See "Microarray"]

Name used to describe arrays of cDNA molecules or oligonucleotides on the surface of microscope slides, membranes, or plastic or silicon chips.

DNA chip [See "Gene chip"; "Genome chip"]

This term is generally used to describe analytical devices for DNA or RNA analysis comprising arrays of oligonucleotides or cDNA on the surface of a silicon, glass, or plastic chip (*5*).

Drexler, K. Eric [See "Nanotechnology"]

K. Eric Drexler (Figure 12) is the author of a series of influential books on nanotechnology, notably, *Engines of Creation* published in 1986, which advocated and explored the possible applications of nanotechnology (*6*).

Figure 12. K. Eric Drexler—nanotechnology pioneer.

Drug discovery

Modern-day drug discovery exploits processes that test thousands of potential drug compounds against thousands of biological targets—the so-called massively parallel screening process. Interactions detected in the screening process provide leads to possible useful pharmaceutical compounds or classes of compounds. The sheer scale of this endeavor and the requirement for rapid low-cost testing necessitate that the screening tests be performed in parallel and that as little sample and reagent as possible be used. Microminiaturization solves these problems by reducing the scale of the screening assay, and providing different types of parallel testing strategies (e.g., microarrays, microwells, nanowells, and micro- and nanobead collections).

Electroendosmosis

Liquids can be made to flow in a microchannel as a result of electroendosmotic forces. This flow arises when solvated cations bound to a negatively charged microchannel wall (charged double layer) move towards the cathode under the influence of an applied voltage (Figure 13). The movement of the ions and the associated water molecules effectively pump fluid along the channel in the direction of the cathode. It is an elegant and practical alternative to mechanical pumps for moving fluid in very small channels.

Electrokinesis

This term encompasses electrical processes used to control movement of fluid and other substances in microchips without the need for valves or pumps. It includes movement of bulk fluids by electroendosmosis, and movement of charged molecules by electrophoresis.

Electronic addressing [See "Electronic hybridization"; "Electronic stringency"]

Electronic addressing is a method for preparing microarrays of biomolecules in a pattern determined by an underlying microelectrode array within a bioelectronic chip. Negatively charged molecules, e.g., biotinylated oligonucleotide probes, can be guided to specific test sites above microelectrodes and then immobilized above the microelectrode. The

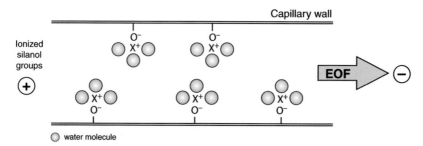

Figure 13. Electroendosmosis-induced flow (EOF) of a fluid in a microchannel. Solvated cations (X^+) move towards the cathode $(-)$, producing a net flow of fluid along the microchannel.

probe solution is introduced into a blank bioelectronic microchip, and the negatively charged probes are drawn to the selected positively charged microelectrode site (all the other electrodes are turned off). Probes are concentrated at the specific electrode and bound to streptavidin in a gel layer overcoating the electrode array. Excess probe is then removed by washing, and the next probe solution introduced into the microchip, and the process repeated at a new microelectrode site. In this way a user-defined array of probes is assembled in a stepwise manner. A unique advantage of this method of producing microarrays is that the microelectrodes can then be used to accelerate and control subsequent hybridization reactions with targets contained in a test solution introduced into the microchip (electronic hybridization and stringency control).

Electronic hybridization

In this process an electrical field is used to manipulate negatively charged oligonucleotide or nucleic acid molecules. Electronic hybridization dramatically speeds up hybridization reactions by as much as 1000-fold compared to conventional diffusion-controlled reactions. One of the hybridization pair, e.g., a DNA probe, is immobilized on an electrode. When the electrode is biased positive, it transports and concentrates the negatively charged DNA target in solution at the electrode surface, thus speeding up the hybridization with the immobilized probe (Figure 14). Reversing the field drives unhybridized or weakly hybridized target away from the immobilized probe on the electrode, and this completes the hybridization process. Adjustment of the electrical field selectively controls dehybridization of probe-bound targets, and permits discrimination of single base mismatches (this process is termed electronic stringency).

Electronic stringency [See "Electronic hybridization"]

Conventionally, ionic strength or temperature is used to ensure complementary binding (perfect match) between a DNA target and a DNA probe. An alternative strategy is to employ electrostatic forces to control hybridization of DNA target to probe molecules im-

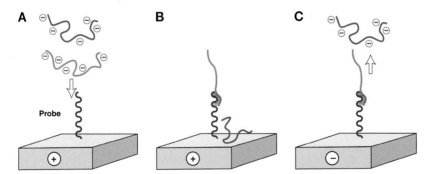

Figure 14. Electronic hybridization: (A) negatively charged target in solution is attracted towards probe immobilized on a positively biased electrode; (B) target hybridizes to the probe; (C) reversing the electrode bias repels non-specifically bound or weakly hybridized molecules away from the probe and the electrode surface.

mobilized on an electrode. A positive bias at the electrode attracts negatively charged target to the surface-bound probes, and reversal of electrode polarity repels weakly or non-specifically bound probes away from the surface into the bulk solution (Figure 14).

Electrostatic bonding [See "Anodic bonding"]

Embossing [See "Wire imprinting"]

A process that uses a tool with a raised pattern that is pressed into a flat surface to create a negative image of the embossed pattern in the surface. Plastic microchips (Figures 15A–15D) can be made this way using a metal embossing tool in a special hot-embossing machine (Figure 15E).

ENIAC [See "Colossus"; "Microcomputers"]

ENIAC is an acronym for electronic numerical integrator and computer. This vacuum-tube-based computer was built at the Moore School of Electrical Engineering, University of Pennsylvania in the United States as part of a joint project with Army Ordnance during World War II. Along with Colossus at Bletchley Park in the UK, this was one of the world's first computers. ENIAC's inception dates back to 1942, and its purpose was to compute ballistic firing tables. Final assembly of ENIAC was completed in 1945, and the machine comprised 30 separate units, 19,000 vacuum tubes, 1500 relays, and hundreds of thousands of capacitors, resistors, and inductors (see http://ftp.arl.mil). On the occasion of the 50th anniversary of ENIAC in 1996, a team from the Moore School of Engineering re-created the computing function of ENIAC on a 0.5-μm CMOS (complementary metal-oxide semiconductor) process single microelectronic chip. The 174,569 transistored ENIAC™-on-a-chip measures 7.44 mm × 5.29 mm and is a fraction of the size of its famous ancestor.

Etching [See "Glass etching"]

A process whereby a design is cut into the surface of a piece of glass or silicon using a dry etch process (e.g., plasma) or a wet chemical etching process. In addition, for crystalline substrates such as silicon, the etch can follow crystal planes (anisotropic etching, e.g., using KOH) or etch all planes equally (isotropic etching, e.g., using a nitric acid + hydrofluoric acid mixture). Conventional etching using photolithography (405 or 436 nm) has a minimum feature size of 0.3 μm, but this can be reduced to 0.1 μm by using X-ray and e-beams to transfer the mask pattern to the surface of the substrate. The new types of microscopes capable of imaging individual atoms and molecules are capable of etching even finer patterns. For example, a scanning tunneling microscope has been used to etch a 150-nm long, 8-nm wide, and less than 1-nm deep groove on the surface of a rubidium molybdenum oxide crystal.

Expressed sequence tags (ESTs)

ESTs are cDNA molecules (sequenced or partially sequenced) that represent expressed genes. They are used on microarrays for gene expression profiling along with

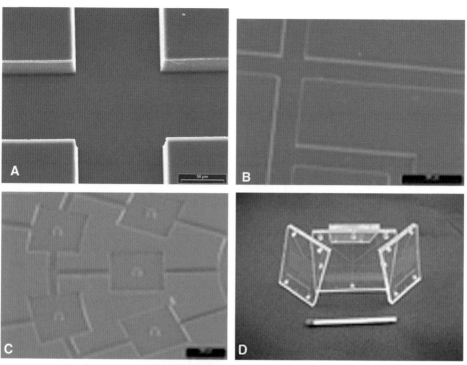

Figure 15. Embossed plastic microstructures and microchips: (A) cross-channel system; (B) multi-channel system; (C) radially disposed reaction chambers and channels; (D) microfluidic chips made using hot-embossed plastic parts (match indicates the relative size of the chips); (E) hot-embossing machine.

Reproduced with permission from Mildendo GmbH, Jena, Germany.

known gene sequences [dbEST lists details of millions of ESTs at http://www.ncbi.nim. nih.gov/dbEST_summary.html].

Expression monitoring

The maxim "DNA makes RNA makes protein" describes the basic process in a biological organism whereby the genetic information encoded in the genes of the cellular genome is converted into specific proteins. In any particular cell, only some of the thousands of genes are expressed, and the degree of expression is variable. Some genes are highly expressed (corresponding to production of a large amount of the corresponding mRNA) whereas others are expressed only at a low level (small amount of corresponding mRNA is produced). The pattern of gene expression can be determined by analyzing the mRNA population using a microarray (the so-called expression monitoring or expression profiling). This can be done directly using mRNA extracted from a cell or more usually by converting the mRNA to cDNA. The cDNA is labeled with a fluorescent molecule (e.g., a cyanine dye, fluorescein or lissamine) and hybridized to the array. The location and intensity of the resulting fluorescent signal on the array correspond to the probe complementary to the cDNA and the amount of that particular cDNA in the sample, and hence the level of expression of that gene. An alternative strategy team-ups up a directly labeled cDNA and a biotinylated cDNA that is detected post-hybridization with streptavidin–phycoerythrin.

cDNA and oligonucleotide microarrays are particularly useful for comparative experiments that analyze the relative levels of gene expression in two different populations of cells or the same type of cell exposed to different conditions (e.g., control cells and cells exposed to a pharmacological agent). The experiment can be conducted on the same array by labeling cDNA from the test cell population with one dye, and labeling cDNA from the control cell population with another dye. The labeled test and control cell samples are mixed together and hybridized to the microarray. Fluorescence scanning of the array reveals signals due to the two different populations of cDNA molecules at individual test sites on the array. Comparison of the signal strengths reveals the relative level of expression of the particular gene in the test and control cell populations.

Femtoliter

One femtoliter is 10^{-15} L (i.e., one-millionth of a billionth of a liter). A typical 12-fluid-ounce beverage can contain 355 mL, which is equivalent to 355 trillion femtoliters of beverage! As a cautionary note: in Great Britain and France a trillion is 1×10^{18} not 1×10^{12} as in the USA, so if you ordered the "trillion fL size" in London or Paris (celui d'un trillion fL), you would get less! It would only be 355 billion femtoliters (and just to add more confusion, a US billion is 1×10^9, but it is 1×10^{12} in Great Britain and France!).

Feynman, Richard Phillips (1918–1988)

Richard Feynman's lecture, entitled "There's plenty of room at the bottom" at the annual meeting of the American Physical Society at the California Institute of Technology, 29 December 1959, is widely acknowledged as a seminal event in the development of nanotechnology (for a transcript go to http://www.zyvex.com/nanotech/feynman.html) (7). He focused attention on the possibilities of manipulating and controlling things on a small scale, the opportunities for design, and the new effects that could exist in a quantum mechanically dominated atomic-scale realm. As an example, he posed the question "Why cannot we write the entire 24 volumes of the Encyclopedia Britannica on the head of a pin?" He then proceeded to show how by a 25,000-fold reduction in the size of the pages it would be possible to fit the entire 24 volumes of the encyclopedia on the head of a pin (Figure 16).

Figure 16. An early nanotechnological challenge—put all 24 volumes of the Encyclopedia Britannica on the head of a pin.

Even the smallest feature, the minute dots in the half-tone reproductions, would occupy the area of 1000 atoms, and so there was still plenty of room for further miniaturization.

He ended his lecture with two $1000 prize challenges. One prize was for the first person making a working electric motor that was only 1/64 in.[3] (this prize was claimed in 1960). The other prize was for the first person writing the information from the page of a book on an area 1/25,000 smaller. This was finally awarded in 1985 to Thomas Newman, who used electron beam lithography to reproduce the first page of Charles Dickens' novel, *A Tale of Two Cities*, on a page measuring 1/160 mm $\times$ 1/160 mm. Subsequently, the Foresight Institute has offered a $250,000 Feynman Grand Prize, in honor of the 1965 Physics Nobel Laureate, for the first person constructing a functional nm-scale robotic arm and a functional nm-scale computing device capable of adding two 8-bit binary numbers (for application details, see http://www.foresight.org).

Fluorescence

This is the light emission that occurs when molecules are raised to an excited electronic state by incident radiation, and then decay to the electronic ground state. Fluorescein is the best-known example of a fluorescent molecule. When a solution of fluorescein is irradiated with light (excitation 494 nm), the fluorescein molecules absorb the energy and are raised to an excited electronic state, and then release light (peak emission at 520 nm) as they decay back to the electronic ground state. Fluorescence is widely used in different types of microchips to signal an event, such as separation of molecular species or intermolecular binding, in or on the microchip.

Functional genomics

Functional genomics expands the scope of biological investigation from studying single genes to studying all genes at once in a systematic fashion. It involves the development and application of genome-wide experimental approaches to assess gene function by making use of the information and reagents provided by structural genomics. It makes use of high-throughput or large-scale experimental methodologies (e.g., DNA microarrays) combined with statistical and computational analysis of the results to understand the link between sequence and function, and to yield new insights into the behavior of biological systems.

Gas chromatograph

In 1975, Terry and Angell built a miniature gas chromatograph that had been micromachined on a 5-cm silicon wafer (8,9). It consisted of a 1.5-m spiral column (20-μm wide × 40-μm deep, coated with OV-101 silicone) capped with a Pyrex glass cover. A diaphragm valve (4-nL internal volume) delivered minute samples into the column for analysis. This was the first example of a microfabricated analyzer. An example of a modern-day micromachined gas chromatograph column, playfully located in a peapod, is shown in Figure 17.

Gene chip [See also "DNA chip"; "Genome chip"]

A "gene chip" is a type of microarray chip that has a surface array of genes—either cDNAs or sets of oligonucleotides representing a gene. It is used to study gene expression or to analyze the sequence of DNA. This type of microchip has also been called a DNA chip, DNA array, genome chip, and genome reader.

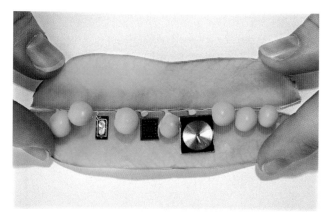

Figure 17. "Lab in a pea pod"—the three microchip devices from left to right are an acoustic wave sensor array, a pre-concentrator for collecting gas vapors for gas-phase analysis, and a gas chromatograph.

Reproduced courtesy of Sandia National Laboratories, SUMMiT* Technologies, www.mems. sandia.gov.

Genome chip

This is a microchip that has attached to its surface the entire set or most of the genes for an organism. The genes are represented by either cDNAs or sets of oligonucleotides. Examples of genome chips based on oligonucleotide arrays include *Arabidopsis*, *Drosophila*, *E. coli*, mouse, rat, and yeast. The human genome is also available on a pair of chips containing over 1,000,000 unique oligonucleotides (25-mers, 11 probe pairs/sequence, 18-μm spot size) representing greater than 33,000 of the best-characterized human genes.

Genomics

This encompasses the analysis of individual genomes and the comparative analysis of genomes from different organisms. Microchips, particularly microarray chips, are playing an important role in genomics as tools for identifying genes and assessing gene expression. More than 70 genomes have been sequenced and sequencing of >400 are currently underway. The complete DNA sequence of the bacterial genomes *Haemophilus influenzae* and *Mycoplasma genitalium* were completed in 1995, and the first complete eukaryotic genome to be sequenced was *Saccharomyces cervisiae* in 1996. This was followed in June 2000 by the announcement of the working draft DNA sequence of the human genome (*10,11*). For information and updates, see the Genomes OnLine Database (GOLD) at http://ergo.integratedgenomics.com/GOLD/ or http://www.tigr.org.

Glass

Glass is an amorphous, solid substance made by fusing silicates with basic oxides (e.g., aluminum oxide, boric oxide). It is a convenient material for capping microchips, and is also used as a substrate that can be etched to make chips. A mixture of hydrofluoric acid and nitric or hydrochloric acid is commonly used for etching glass. HF reacts with glass according to the following equation:

$$SiO_2 + 4HF \rightarrow SiF_4 + 2H_2O$$

The purpose of the added nitric or hydrochloric acid is to solubilize metal fluorides, originating from metals present in glass. The glass is masked for etching with an etch-resistant layer of amorphous silicon, or with a sandwich consisting of a chromium adhesion layer and a gold protection layer.

Gridding

Synonymous with arraying, gridding refers to the process of forming an array of samples or reagents. It was originally used in the context of a microbial enumeration assay based on a square grid that was printed onto a cellulose ester membrane using silicone grease or hydrocarbon wax (*12*).

High-throughput [See "Drug discovery"]

All three main types of microchip have found a role in the current need for high-throughput testing. This type of testing is being driven by the drug discovery programs that require a very large number of assays as part of the screening procedures to identify potentially new drug molecules.

Hot piranha

Not something straight from the fish pot; instead a dangerous, highly toxic and corrosive liquid, much loved by the microfabrication community for cleaning silicon. The recipe for this fatal fluid is three parts 98% sulfuric acid and one part 30% hydrogen peroxide by volume.

Hybridization [See "Microarray"]

The formation of a partially or completely double-stranded (duplex) nucleic acid formed by the base-pairing (A is complementary to T and G is complementary to C) of two single-stranded nucleic acids (DNA or RNA) or two oligonucleotide probes, or two deoxyoligonucleotide probes, or combinations of these molecules is termed hybridization. Sometimes it is referred to as annealing. This reaction involves hydrogen bond formation between the complementary bases on the different strands of the duplex (two H-bonds for A:T and three for G:C). Hybridization can occur in solution, or with one of the reactants immobilized on a substrate. This reaction underlies all the applications of cDNA and oligonucleotide microarrays. Hybridization reactions are strongly influenced by temperature and ionic strength. Elevated temperature and low ionic strength favor highly specific hybridization (perfect match) over reactions in which some of the bases in the hybridized molecules are not complementary (mismatch). At higher temperatures the hybrids dissociate, and this is often termed "melting" or "denaturation." The temperature and buffer ionic strength control the fidelity of hybridization and allow discrimination between a perfectly complementary hybridization (perfect match) and a mismatch by as little as a single base.

Pairs of probes designed as a perfect match and a mismatch for the intended target are used on oligonucleotide microarrays to assess the specificity of hybridization. A positive result at the perfect match oligonucleotide is reliable only if the signal from the mismatch probe site is negative.

Hybridization chip [See "SBH"]

Imprinting

Wire imprinting is a simple method to fabricate microchannels or other features in microchips. A wire framework (wire diameter ~80 μm) that corresponds to the desired channel design is constructed. This tool is then pressed into the surface of a piece of plastic to create the channels as surface indentations with the same geometry as the wire tool.

Injection molding

This popular fabrication process employs a multi-piece mold into which molten plastic is injected. The plastic adopts the shape of the mold, and after it solidifies and cools, the mold is separated to release the plastic injection molded object. Injection molding can be adapted to produce microchip devices and has been successfully used to make 7 × 7 arrays of 1-μL volume wells for real-time PCR.

Ink-jet

A drop-on-demand ink-jet is a device used in printers to deliver small droplets of ink onto a two-dimensional surface. Drops are formed by means of a piezoelectric, thermal, or acoustic pressure generator that ejects ink through a small orifice. Droplet sizes as small as 1 μm in diameter can be attained. The ink-jets can be combined into multi-jet print heads for higher throughput printing (Figure 18A). The ink is replaced by a biological reagent (e.g.,

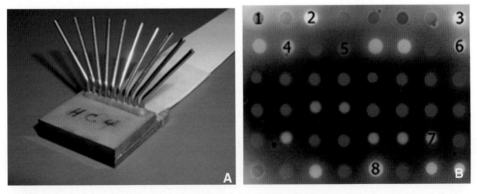

Figure 18. (A) A 10-fluid integrated piezoelectric drop-on-demand array print head and (B) a microarray of eight different dyes (labeled 1–8) printed onto 200-μm centers using this print head.

Reproduced with permission from MicroFab Technologies, Inc., Plano, TX.

oligonucleotides) and used to print microarrays directly onto the surface of a glass slide (Figure 18B) or chemical reagents for in situ synthesis of oligonucleotide microarrays.

In situ synthesis [See "Light-directed synthesis"]

This is an example of a bottom-up fabrication process for microchips. It is mainly used to construct oligonucleotide arrays. In situ synthesis of oligonucleotides has been achieved by classical DNA synthesis chemistry and by means of light-directed methods. The general process consists of the following steps:

i. React the surface hydroxyl groups on a silicon or glass substrate surface with aminopropyl silane.
ii. Attach a linker (e.g., hexaethyleneglycol) to the surface amine groups. The end of the linker then serves as the starting point for standard phosphoramidite or phosphonate DNA synthesis using side group-blocked bases protected with either acid-labile (e.g., dimethoxyltrityl) or photolabile [e.g., nitroveratryloxycarbonyl (NVOC)] protection groups.
iii. Synthesize oligonucleotides in a stepwise manner at precise locations on the surface. This is achieved in one of the several ways: (i) apply the reagents as minute drops using an ink-jet printer, (ii) flow the reagents over specific areas of the surface defined using a physical barrier, and (iii) direct light to specific surface locations using a photolithographic mask or a micromirror array.

Integration [See "i-STAT analyzer"]

A key advantage of microminiaturization and microchips is the option to integrate a series of sequential analytical steps onto a single device. In this way a time-consuming and complicated analytical procedure can be simplified and completed sooner. Examples of analytical processes that have been integrated onto a single analytical microchip include cell isolation + PCR, immunoassay + detection of bound and free fractions, sample preparation + PCR + capillary electrophoretic analysis of the PCR reaction mixture, and PCR + real-time analysis of amplicon formation using a Taq-Man™ assay.

A series of different microchips can be combined into a single analyzer as illustrated by the hand-held microchip instrument built at Sandia National Laboratories (http://www.sandia.gov) (Figure 19A). It incorporates sample concentration, acoustic wave sensor array, and gas chromatograph microchips, and was designed to detect explosives (e.g., sniff-out landmines). A similar degree of integration is achieved in the PCR system shown in Figures 19B–19D. This device processes the sample in a microfluidic device that fits into a thin flat miniature plastic PCR tube. The PCR tube is moved to a small instrument that performs both the thermocycling and quantitates amplicons produced in the PCR reaction.

Ion-channel microchip

Patch-clamp assays can conveniently be performed in specially designed bioelectronic microchips (Figure 20A). The key microfabricated feature is an ~1.2-μm diameter hole at the base of a microchamber etched into a membrane separating two microchambers (Figures 20B and 20C). Microelectrodes adjacent to the hole capture the cell, which is then

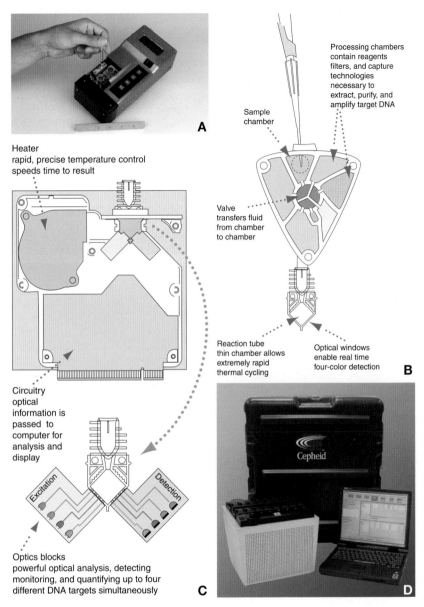

Figure 19. Integrated microchip analyzers: (A) hand-held portable chemical analysis-on-a-chip device for testing liquids or gas mixtures. A minute sample (0.1 nL) is analyzed by the on-board acoustic wave sensor array, gas vapor pre-concentrator, and gas chromatograph; (B) schematic representation of the Cepheid microfluidic processing module inserted into a reaction tube; (C) reaction tube inserted into the I-CORE® (Intelligent Cooling/Heating Optical System) temperature-controlled fluorimeter for performing and continuously monitoring PCR reactions; (D) SmartCycler® TD System for use with the I-CORE® module (16 programmable reaction sites, rugged travel case and laptop computer).

Reproduced courtesy of Sandia National Laboratories, SUMMiT* Technologies, www.mems.sandia. gov (Figure 19A); Reproduced with permission from Cepheid, Sunnyvale, CA (Figures 19B–19D).

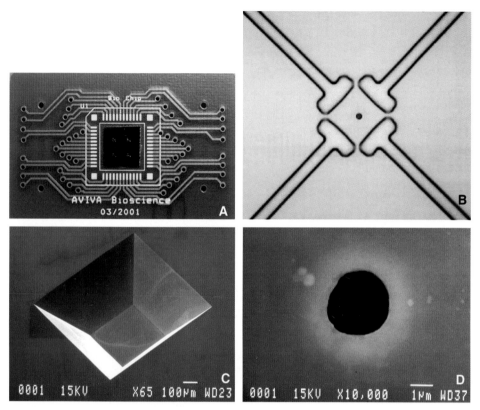

Figure 20. Bioelectronic chip for "patch-clamp" ion-channel analysis: (A) bioelectronic chip on connection board; (B) electrodes for manipulating cells into the micromachined well; (C) close-up of the micromachined well; (D) close-up of micropore (~1.5 μm in diameter) at base of well where a single cell is positioned by suction for analysis.

Reproduced with permission from AVIVA Biosystems Corporation, San Diego, CA.

firmly wedged into the hole by applying suction to the lower chamber of the microchip (Figure 20D). The part of the cell membrane sucked into the hole is then ruptured in order to establish a connection between the upper and lower chambers. A test agent is added to the upper chamber, and the response of the ion channels in the cell wall is assessed from electrical measurements by electrodes located in the two chambers.

IP [See "Controversy"]

Popular abbreviation for intellectual property—patents, trademarks, and copyright material. It is the source of much controversy and conflict in the microchip arena.

i-STAT analyzer

The i-STAT analyzer represents one of the earliest commercial microchip devices (Figure 21A). It comprises a cartridge containing an analyte-sensitive silicon microchip, and a

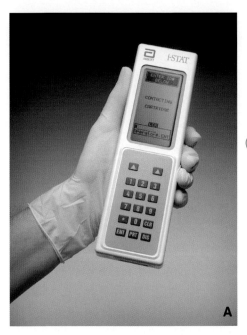

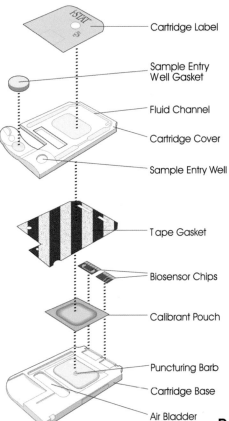

Cartridge Label

Sample Entry
Well Gasket

Fluid Channel

Cartridge Cover

Sample Entry Well

Tape Gasket

Biosensor Chips

Calibrant Pouch

Puncturing Barb

Cartridge Base

Air Bladder

D

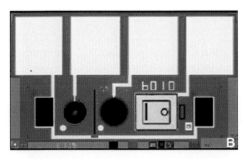

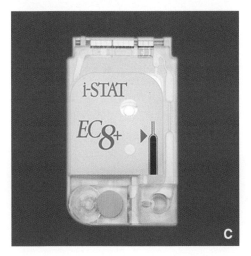

Figure 21. i-STAT® Portable Clinical Analyzer. The hand-held battery-operated device (A) analyzes whole blood samples for a range of common analytes using silicon microchip-based biosensors (upper row of structures) (B), housed in a cartridge (C), shown in exploded view in panel (D).

Reproduced with permission from i-STAT Corporation, East Windsor, NJ.

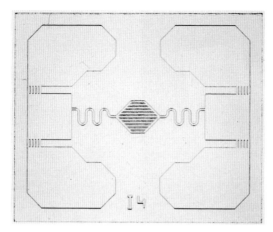

Figure 22. In vitro fertilization microchip. The etched 15-mm × 17-mm glass chip comprises a grooved central hexagonal semen chamber linked to the square egg chambers at the ends of the two tortuous channels (100-μm wide × 40-μm deep). Each egg chamber is linked to two reservoirs to minimize evaporation during prolonged incubations. The microchip is capped with a glass top with access ports above the semen and egg chambers (not shown). Semen is applied to the semen chamber and sperm swims along the channel into the egg chambers and fertilize eggs. The channels present an obstacle such that the fastest swimming sperm (these have the highest fertilizing potential) arrive in the egg chambers first, and have the best chance of fertilizing the eggs.

hand-held analyzer. The chip contains minute electrochemical sensors (e.g., for sodium, potassium, glucose) constructed from layers of immobilized analyte-sensitive reagents (e.g., a layer of glucose oxidase for the measurement of glucose) (Figures 21B–21D). Blood is pipetted into the cartridge, and the cartridge inserted into the analyzer. An automatic sequence ensues in which first the calibrator, and then the blood sample, flows over the sensors. The analyzer measures the signals produced by the chips in response to the analytes in the sample, and the final results are displayed on the screen built into the analyzer (Figure 21A).

IVF chip

An in vitro fertilization (IVF) microchip, comprising two open chambers connected by a microchannel, provides a novel means of selective fertilization of eggs (Figure 22). Semen is placed in one chamber and eggs in the other. Sperm swims from the semen chamber through the microchannel into the egg chamber where they fertilize the eggs. The channel has a series of twists and turns, and this selects the fast swimming sperm that have the greatest fertilizing potential.

Jokes

Microchips are a serious subject, but you may be interested to know that you can buy "Joke Chips," "JokePicker Chips," and "Laugh Chips," all as part of "Funny Bits from Your Talking Chips™" (see http://www.funnybits.com)!

Knowledge

"A little knowledge is a dangerous thing" (Alexander Pope, 1688–1744). Somehow this quotation seemed appropriate for this book, bearing in mind its topic!

Lab-on-a-chip

A new journal published by the Royal Society of Chemistry devoted to the "chemical, biological, and engineering aspects of lab-on-a-chip technology and its applications" (see www.rsc.org/loc).

Laboratory on a chip

The phrase "laboratory on a chip" was originally used to describe virtual experiments conducted on a computer *(13)*. The "chip" was the microelectronic chip in the computer, and the "laboratory" was the virtual realm created by the software program. Subsequently, in the chemical analysis field, "lab-on-a-chip" or "lab-on-chip" has been used to describe a self-contained microchip that would perform all the steps in an analytical process *(14)*.

Laser capture microdissection

This is a powerful new technique that allows precise selection of cells in a tissue section and specific removal of the selected cells. These cells can then be subject to analysis using, e.g., microarray chips. The process utilizes a light-sensitive film placed in contact with the tissue section. The cells are located using a microscope, and a burst of laser light dissects the cells, which then adhere to the film. The film, with the selected cells adhering to it, is pulled away from the tissue section for subsequent processing and analysis of the captured cells, e.g., using a microarray.

Laser drilling

In the microchip fabrication process, it is often necessary to provide small holes through chip components that can serve as entry or exit ports. Laser drilling is one of the number of ways to drill holes into chip components. A CO_2 laser drills via a combination of melting, ablation, and vaporization at the point where the laser impinges on the surface of the chip. Hole sizes less than 200 μm are possible with laser drilling.

LIGA

Lithographie Galvanoformung Abformung is a micromolding process used to construct microdevices with high aspect ratios. LIGA consists of a series of steps. A design is transferred with X-rays through a mask (X-ray lithography) to a polymethylmethacrylate (PMMA) resist coated on a base plate. The resist is then developed to create a high aspect ratio mold. A metal (e.g., nickel) is then deposited into the mold in an electroforming process to produce a complementary structure. This can be the final product, or the electroplated part can serve as a mold for casting further structures.

Light-directed synthesis

The light-directed synthesis method devised by Fodor and colleagues at Affymetrix (also known as VLSIPS™) is used to construct oligonucleotide arrays and peptide arrays.

The steps in this in situ synthetic method for an oligonucleotide are as follows (Figure 23):

i. Surface hydroxyl groups on a silicon or glass substrate are reacted with aminopropyl triethoxysilane.
ii. A hexaethyleneglycol linker functionalized with a light-sensitive group (e.g., NVOC) is reacted with the free surface amino groups.
iii. Exposure of the surface to light in a pattern defined by a photomask initiates a photochemical deprotection reaction. This generates reactive hydroxyl groups, and these then become the site for reaction with a 3'-O-activated 5'-O-photoprotected deoxynucleoside.
iv. Successive rounds of light-directed deprotection and subsequent reaction with 3'-O-activated 5'-O-photoprotected deoxynucleosides facilitate base-by-base synthesis of oligonucleotides at precisely defined μm-sized locations. Usually, oligonucleotides up to 25-mers are produced at sites measuring less than 50 μm × 50 μm, and densities of more than 10^6 sites/cm^2 are possible.

Lilliputian

An apt description for microchips, the tiny inhabitants of tomorrow's analytical world.

Linker

A linker molecule is used in microarray chip manufacture to space apart the arrayed molecule from the chip surface. This makes the arrayed molecule more accessible to reactants in samples applied to the chip surface by minimizing steric hindrance effects due to the chip surface. Hexaethyleneglycol molecules, for example, are used to space an oligonucleotide away from the surface of a glass chip or silicon chip.

Liquid chromatography

A liquid chromatograph has been microfabricated on a 5-mm × 5-mm × 0.4-mm silicon chip. It consists of a 15-cm long spiral separation column and a 1.2-pL detection cell. This is just one example of the many classical analytical instruments and techniques reduced to fit onto a single chip.

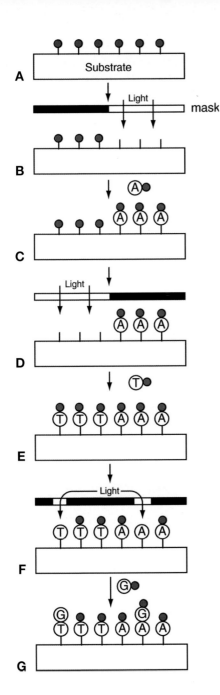

Figure 23. In situ light-directed synthesis of an oligonucleotide microarray: (A) surface hydroxyl groups on a silicon or glass substrate are activated by reaction with aminopropyl silane and a linker (e.g., hexaethyleneglycol) derivatized with a photolabile group (●-); (B) illumination of the surface through a mask generates reactive groups on the end of the linker; (C) surface groups serve as the starting point for standard phosphoramidite or phosphonate DNA synthesis using side group-blocked bases protected with a photolabile (e.g., NVOC) protection group (A-●, T-●, G-●, C-●); (D–G) successive rounds of deprotection using different masks and reaction with protected bases builds oligonucleotides on the surface of the substrate at the locations determined by the mask patterns.

List server

A convenient way for a global community to share information and communicate is via a list serve. Electronic mail sent to the list is copied and then automatically sent to all subscribers. For a gene array list serve, sign up at http://www.bsi.vt.edu/ralscher/gridit/

g_alistserv.htm; or for both gene array and microarrays sign up at Gene-arrays@itssrv1. ucsf.edu; and for the Stanford MicroArray forum, visit the Brown Lab's site complete guide for microarraying for the molecular biologist and select MicroArray forum (http://cmgm. stanford.edu/pbrown/mguide/index.html).

Lithography [See "Photolithography"]

LOC

Abbreviation for lab-on-a-chip. Also, in the clinical laboratory context, it stands for "left on cells," a common pre-analytical disaster that befalls some blood specimens.

μL

Abbreviation for a microliter (one-millionth of a liter). One drop is 1/60th of a teaspoon, and this is 1/200th of a liter, so a drop is 83.3 μL! (see http://allmeasures.com/).

μm

Abbreviation for a micrometer (one-millionth of a meter). A light year is equivalent to 9,460,900,000,000,000,000,000 μm.

μ-TAS

An abbreviation for "miniaturized total analytical system" coined by Andreas Manz and colleagues (15) (Figure 24). It denotes a microminiaturized device that integrates all the components and steps in an analytical procedure (sampling, sample handling, chemical reactions, separations, detection) into a small chip-sized device, i.e., a laboratory shrunk to fit on a very small area the size of a chip.

Mask

A mask is a template that determines the pattern of reaction, deposition, or etching on the surface of a substrate. The microchip design to be etched is transferred to the photoresist-

Figure 24. Andreas Manz—μTAS pioneer.

Figure 25. Chrome-on-glass masks for transferring different microchip designs to photoresist-coated wafers.

coated silicon or glass substrate surface by projecting the design onto the surface, or by placing the mask in contact with the surface. An example of two chrome-on-glass masks used in microchip manufacture is illustrated in Figure 25. Depending on the structure and complexity of the microchip design, a series of masks may be used in the fabrication process. Each mask defines a particular part or layer of the structure. After each completed etch step, the wafer is recoated with photoresist and exposed using the next mask in the series.

Maskless synthesis [See "Digital light processing"]

Mass spectrometry chips

A miniature mass spectrometer has been fabricated using the LIGA process. Micromachined accessories for mass spectrometry include arrays of microfabricated 15-µm diameter nozzles for converting liquid sample emerging from a liquid chromatograph into gaseous ions at atmospheric pressure for injection into mass spectrometers (electrospray ionization) (Figure 26). The total volume of these devices is less than 25 pL, and they can achieve nL/min flow rates.

MEMS [See "Digital light processing"]

A microelectromechanical system (MEMS) is a hybrid device that contains mechanical elements, sensors, actuators, optical components, and electronics. They are fabricated using a combination of integrated circuit technologies (oxidation, diffusion, photolithography, sputtering) and micromachining technologies (bulk, surface and sacrificial layer micromachining, LIGA, reactive ion etching, micromolding, bonding). MEMS devices are already in use to sense deceleration and trigger airbag release in automobiles (MEMS accelerometers), and to control light in video projection systems (DLP chips). Some examples of the scope and intricacy of micromachined devices are illustrated in Figure 26.

Mesoscale

Derived from the Greek *mesos* for middle, this is the scale above molecular but below macroscopic.

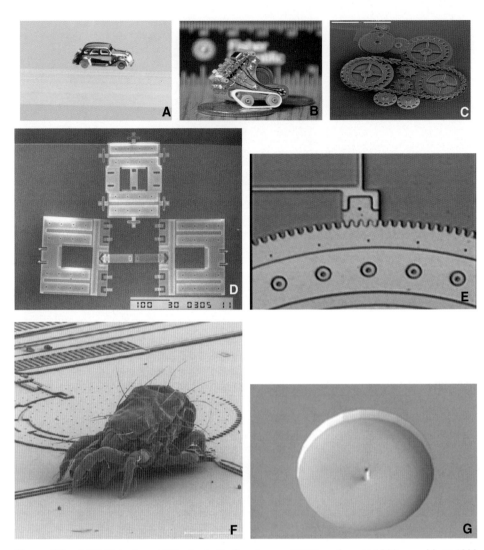

Figure 26. MEMS devices: (A) body shell of a microcar (4800 μm long × 1800 μm wide × 1800 μm high); (B) A 1/4 in.³ autonomous untethered robot that turns on a dime and parks on a nickel; (C) silicon microchain device (demonstrates engaging-device drive gears); (D) multi-part hinged structure; (E) gear teeth on a wheel interacting with stepper teeth on an *X–Y* actuator; (F) spider mite on a microlock; (G) one of the nozzles (26-μm in diameter) from the 100-nozzle array ESI Chip™ for nanoelectrospray mass spectrometry.

Reproduced with permission from DENSO Corporation, Kariya, Japan (Figure 26A); courtesy of Sandia National Laboratories, SUMMiT* Technologies, www.mems.sandia.gov (Figures 26B, 26C, and 26F); © ZYVEX Corporation, Richardson, TX (Figures 26D and 26E); Advion BioSciences, Inc., Ithaca, NY (Figure 26G).

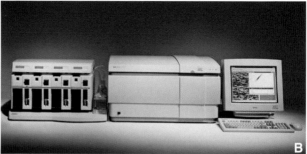

Figure 27. Microarray chips: (A) Affymetrix GeneChip® probe array; (B) the GeneChip® Instrument System.

Images courtesy of Affymetrix, Inc., Santa Clara, CA.

Microarray [See "Array"; "Electronic addressing"]

This usually refers to an array of reagents (oligonucleotides, cDNA, peptides, proteins) synthesized in situ or spotted on a small chip or a microscope slide, or immobilized on a microelectrode array. Depending on the arrayed material, these devices are known as gene chips, genome chips, tissue microarrays, or protein chips. Microarrays have emerged as important analytical tools in gene expression monitoring and drug discovery. There is a growing selection of commercially available microarrays (Figure 27A) and associated analytical equipment (e.g., array scanners) (Figure 27B). Many laboratories have chosen the spotting technique as a way to make microarrays, using one of the many spotting machines available commercially. Very high densities of arrayed compounds are possible (e.g., densities of greater than 1 million oligonucleotides/mm^2). Analysis at this scale produces massive amounts of data and has necessitated a strong emphasis on bioinformatics techniques. If you are interested in the data aspect of microarray analysis, you can view databases of raw and normalized data from microarray experiments, and the corresponding image files at http://genome-www5.stanford.edu/MicroArray/SMD/.

Microarray spotter [See "Arrayer"]

An instrument that deposits μL–nL volumes of reagent solutions at precisely controlled locations on the surface of a substrate to produce an array (Figure 28A). The process is usually conducted in an enclosed chamber. Reagents are held in microwell plates, and a robotic arm carrying a printing head (e.g., a pin array) (Figure 28B) picks up and spots reagents onto slides laid out inside the chamber.

Microcantilever

Silicon-micromachined cantilevers (microcantilevers) have established applications in atomic force microscopy (Figure 29A), but new microcantilever-based assays are emerging. A microcantilever (200-μm long × 50-μm wide × 2-μm thick) (Figure 29B) coated

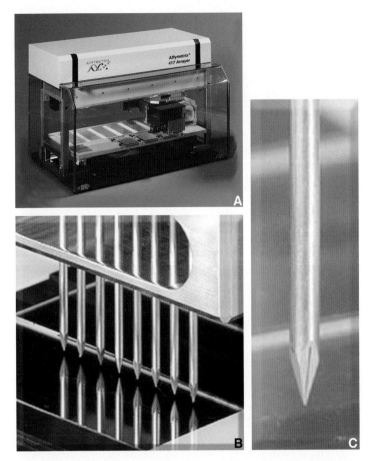

Figure 28. Spotters for making microarrays: (A) the Affymetrix 417™ Arrayer; (B) Quill pin spotting head; (C) close-up of a pin (75- to 200-μm spot size, 100 spots/loading of the pin).

Images courtesy of Affymetrix, Inc., Santa Clara, CA (Figure 28A); Cartesian Technologies, Inc., Irvine, CA (Figure 28B and 28C).

with a binding agent (e.g., antibody) will bend in response to the increased mass caused by capture of specific antigen (e.g., prostate-specific antigen). The minute deflection of the cantilever can be detected optically and related to concentration of the analyte in the test solution. Arrays of cantilevers each coated with a different capture agent offer an alternative microchip-based strategy for simultaneous multi-analyte assays.

Microchannel

A channel having μm-size in at least one dimension (length, width, or depth). Microchannels are important microchip components that are used to transfer fluid within microchips and for separations (e.g., capillary electrophoresis).

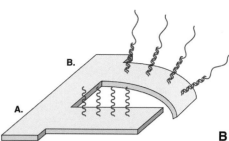

Figure 29. Microcantilevers. (A) Example of a 160-μm long × 50-μm wide silicon microcantilever with a tetrahedral tip for atomic force microscopy; (B) schematic representation of a microcantilever-based assay: the probe-coated cantilever **A** hybridizes with specific target **B** and bends as a consequence of the increased mass of the cantilever.

Reproduced with permission of Olympus Optical Company Ltd., Japan (Figure 29A).

Microchip [See "Lab-on-a-chip"; "Microminiaturization"; "ITAS"]

A tiny square of a silicon semiconductor wafer that is etched and doped to form a large number of circuit components (transistors, diodes, capacitors, resistors) that make up an integrated circuit.

The meaning of the word "microchip" has been extended to include different types of analytical microchips, and I like to divide analytical microchips into three broad categories: microfluidic chips, bioelectronic chips, and microarray chips.

Microfluidic chips contain microchambers and microchannels designed for fluidic manipulation of liquid samples. Bioelectronic chips contain microelectronic components (e.g., microelectrodes) within microfluidic structures. Microarray chips contain reagents immobilized as microarrays, and these can be on a surface or enclosed within a microfluidic compartment. A range of fabrication techniques have been developed to construct analytical microchips including etching, embossing, imprinting, printing, and in situ synthesis.

Microchips have been designed for numerous analytical applications, and representative examples are listed in Table 1. In addition, different steps in an overall analytical process can be integrated on a single microchip device to produce a "lab-on-a-chip" or a μTAS.

Microcomponents

Many everyday objects and devices have been miniaturized and these provide a wealth of components for the microchip designer (Table 2).

Microcomputer

Miniaturization in electronics has successively transformed computers from mainframe behemoths, to relatively large racked systems, to smaller desktop, and still smaller

Table 1.   Analytical and Preparative Applications of Microfluidic Chips, Bioelectronic Chips, and Microarray Chips

Blood analysis (glucose, urea, hematocrit, etc.)	Ion channel assay
Blood typing	Isotachophoresis
Capillary electrophoresis (DNA, RNA, protein sizing)	Ligase chain reaction (LCR)
	Liquid chromatography
Cell lysis	Micro-organism isolation
Cell filtration	mRNA analysis
Cell fusion (IVF)	Nucleic acid hybridization (DNA, RNA, oligonucleotides)
Cell isolation (red vs white cells)	
Cell selection (rare cells)	Immunoassay
Cell sorting	Liquid chromatography
Coagulation testing	Mass spectrometry
DNA sizing	Polymerase chain reaction (PCR)
Degenerate oligonucleotide polymerase chain reaction (DOP-PCR)	Protein sizing
	Restriction fragment length polymorphism (RFLP) analysis
Electrolyte analysis	
Expression analysis	Sample delivery (electrospray ionization)
Field-flow fractionation	Semen analysis (motility, vitality, penetration)
Immunoassay	

laptop computers, into today's hand-held personal digital assistants. A similar trend can be seen in the clinical laboratory where microelectronics has shrunk analyzers to the hand-held scale. Developments in microfabrication and nanotechnology promise even smaller, and more sophisticated, microanalyzers in the future. The original vacuum-tube-based computers, ENIAC, at the Moore School of Electrical Engineering, University of Pennsylvania in the United States, and Colossus at Bletchley Park in UK, needed entire rooms, as is clear

Table 2.   Microchip Components

Accelerometer	Interferometer	Refrigerator
Air turbine	Ion-specific sensor	Relays
Analyte-specific sensor	Laser	Solenoids
Beam	Lever	Spectrophotometer
Bearing	Light-emitting diode	Strain gauge
Cantilever	Mass spectrometer	Switch
Computer chip	Motor	Thermal detector
Flow meter	Microphone	Tuning forks
Fluidic amplifier	Mirror array	Turbine
Gas chromatograph	Particle filter	Tweezers
Gears	Positioning systems	Vacuum gauge
Grippers	Pressure sensor	Vacuum tube
Heat exchanger	Pressure switch	Valve
Incandescent light	Print heads	Voltage sensors
Ink-jet	Pump	

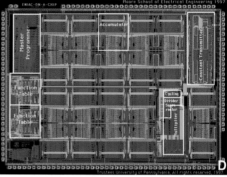

Figure 30. Computers and microprocessors: (A, B) ENIAC at the Moore School of Electrical Engineering, University of Pennsylvania, Philadelphia, PA; (C) Colossus at Bletchley Park, UK; (D) ENIAC™-on-a-chip (7.44 mm × 5.29 mm); (E) modern-day Intel® Pentium® 4 Processor (0.13- and 0.18-μm manufacturing process; speeds up to 2.2 GHz) and Intel® 850 chip set.

Reproduced with permission from John M. Mauchly Papers, Rare Books & Manuscripts Library, University of Pennsylvania, Philadelphia, PA (Figures 30A and 30B); Tony Sale (Figure 30C); Jan van der Spiegel, Electrical Engineering, University of Pennsylvania, Philadelphia, PA (Figure 30D); Intel® Corporation, Santa Clara, CA (Figure 30E).

from the early photographs reproduced in Figures 30A–30C. On the occasion of the 50th anniversary of ENIAC in 1996, the computing function of this machine was re-created using a 0.5-μm CMOS process on a single microelectronic chip. The ENIAC™-on-a-chip (Figure 30D) measures 7.44 mm × 5.29 mm and has 174,569 transistors. A modern-day microprocessor chip illustrates the immense strides that have been made in miniaturization in electronics since the pioneering days of the ENIAC and Colossus (Figure 30E).

Microcontact printing [See "SAMs"]

Surface μm- and nm-sized structures can be made by microcontact printing. First, a poly(dimethylsiloxane) (PDMS) stamp that bears the required design is made. It is inked with a thiolate, and applied to the surface of a gold substrate. Raised portions of the stamp transfer thiolate to the gold surface to produce the intended design.

Microdispensing

Different spotting and printing devices have been developed to produce microarrays of immobilized reagents. They are also used to dispense micro- and nano-volumes of sample or reagent into analytical microchips.

Microelectrode array [See "Bioelectronic microchip"]

Arrays of microelectrodes of varying geometries and complexity have been fabricated within microchips (bioelectronic microchips) for cell isolation and manipulation. Some illustrative examples of microelectrode structures inside bioelectronic microchips are given in Figures 31A–31C.

Microfiltration [See "Cell isolation"]

Simple arrangements of posts are effective for filtering cells in microchannels, as shown in Figure 8. Alternatively, a weir structure shown in Figure 9 can isolate cells by entrapment in the μm-wide gap between the top of the weir and the underside of the glass top capping the microchannel on the chip.

Microfluidics [See "Electroendosmosis"; "Electrokinesis"]

This is the study of the flow behavior of microliter and sub-microliter volumes of fluids in μm-sized channels and chambers. Microfluidic modeling provides a convenient

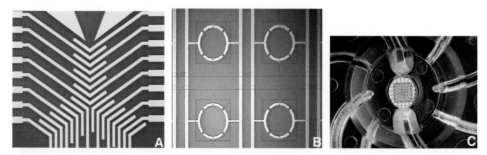

Figure 31. Bioelectronic microchips containing microelectrode arrays. (A) Particle switch, (B) cell positioning bioelectronic chips, and (C) microelectrode array for DNA probe localization (NanoChip™ Nanogen, Inc.).

Reproduced with permission from AVIVA Biosystems Corporation, San Diego, CA (Figures 31A and 31B).

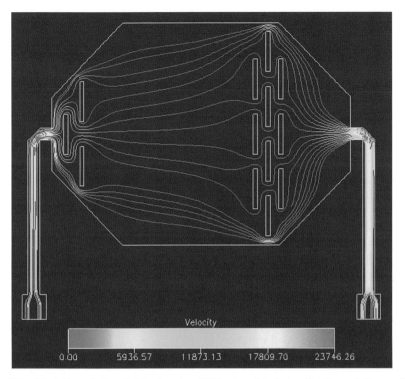

Figure 32. Modeling of microfluidic flow in a PCR microchip [microfluidic modeling performed using CoventorWare™ ANALYZER™ software package (Coventor, Inc., Cary, NC)].

tool for intelligent microchip design. Typical results of the predicted flow pattern in a combined cell filtration–PCR microchip are illustrated in Figure 32.

Microliter

One microliter is 10^{-6} L (i.e., one-millionth of a liter). As an indication of just how little this is, a British Imperial pint of beer (34.678 in.^{3}) contains 567,701 µL. *Caveat emptor*—the same pint in the US would contain only 473,176 µL, because the US pint is smaller (28.875 in.^{3})!

Micromachining

Physical and chemical treatment of a solid substrate (silicon, glass, quartz, plastic) using photolithography, etching, ablation, film deposition, and bonding to produce a microsized device. Some of the wide range of micromachinable materials are listed in Table 3.

Micrometer

One-millionth of a meter (10^{-6} m). The Empire State building is 1252 ft tall—equivalent to 381 million µm!

Table 3. Materials for Micromachining

Alumina	Gallium arsenide	Schott B270 glass
Aluminum	Gold	Silicon
Ceramics	Indium phosphide	Silicon carbide
Copper	Pyrex glass	Soda glass
Diamond	Quartz	Tempax glass
Fluorocarbon polymers	Rubidium molybdenum oxide	

Microminiaturization [See "Laboratory on a chip"; "μ-TAS"]

Analytical processes and analyzers miniaturized to the μm- or even the nm-scale have a number of advantages.

Ease of manufacture—Most devices can be made using the high-volume manufacturing processes in everyday use in the microelectronics or printing industry.

Rapid design cycle—A large number of different designs can be produced simultaneously on one silicon wafer, facilitating rapid prototyping and rapid design optimization.

Reduced operating costs—Reagent consumption in devices with total volumes less than 1 μL is very low, and this leads to cost reduction compared with conventional analyzers.

Portability—Microchip analyzers are small and lightweight, and so can be readily used at the point-of-care (e.g., bedside, doctor's office) or in extra-laboratory applications (e.g., streamside).

Rapid analysis—Microchip-based analyzers can operate quicker than their macroscale counterparts because of the reduced scale (e.g., diffusion distances). Also, complex assays can be completed sooner when the process is integrated onto a microchip.

High-throughput multi-analyte assays—The microarray format provides thousands of test sites on a single microchip so that thousands of different analyses can be conducted simultaneously.

Sample size—The volume of both sample and reagent is significantly reduced in microchip-based analyzers (e.g., nL or pL volumes of sample) and this can be critical in analytical situations in which the amount of a specimen is limited (e.g., forensic samples).

Disposal and safety—The sub-microliter sample volumes commonly used in microchips reduce the exposure of the analyst to hazardous samples and minimize the volume of waste fluids. Also, the minute size of the entry ports into microchips makes possible entombment of the contents of a microchip (unused sample and reagents, reaction mixtures) for safe disposal.

Reliability—Multiple identical test sites on the same chip permit a cost-effective means of performing replicate analyses and hence improving the reliability of an analytical result.

System integration—All the components of an analyzer can be integrated onto the microchip by microscale fabrication. This makes possible integration of the various steps in an analytical process, from sample addition to outputting the result, in an economical way not possible in the macroworld.

Micromirror [See "DLP"]

A component of a digital light processing (DLP) chip. Each mirror measures 16 μm × 16 μm, and is mounted on a hinge so that it can tip 10°, and so control the direction of reflected light.

Micron

Also known as micrometer (10^{-6} m).

Microneedles

Just to make the point (!) about the scope and diversity of microfabrication, I have included this entry on microneedles. These are the syringes of the future. Very small microneedles (e.g., 150-μm long) penetrate only the outermost layer of the skin where there are no nerve endings, so no pain is felt when drugs or vaccines are administered with these devices!

Micropatterning

[See "Microcontact printing"; "SAMs"]

Microphysiometer

The microphysiometer was among the earliest examples of a micromachined analytical device. One type comprises a 1-mm^2 array of 50-μm wide × 50-μm deep pH-sensitive microwells (125-pL volume) etched in silicon and capped with a glass cover slip to form a fluid channel. Cells are loaded into the wells and the cellular response to different compounds is assessed from measurements of the acidity of the culture medium. The pH of the culture medium is determined potentiometrically by means of a light addressable potentiometric sensor (LAPS) built into each well. This measures the surface potential in regions illuminated by light-emitting diodes mounted underneath the silicon chip, and depends on the pH of the fluid in the microwell.

Microscope slide

The familiar 1-in. × 3-in. glass microscope slide has become a popular substrate for printing microarrays. The surface of the slide is often treated with poly-L-lysine or an aminoalkylsilane to make it positively charged in order to improve binding to negatively charged cDNA or oligonucleotide molecules.

Microspot immunoassay

Arrays of spots (50 μm^2) containing antigens or antibodies can be produced and used for simultaneous multi-analyte immunoassays. This type of immunosensor (also termed a microanalytical "compact disk") can be used to test a sample for millions of different analytes on a 1-cm^2 chip, and is an early example of a protein chip (16).

Microtransponder

One of the most sophisticated bioelectronic chips built so far is the microtransponder device, shown in Figure 5D. It can be used to detect and differentiate large numbers of DNA targets in a single assay. The device is an oligonucleotide probe-coated integrated circuit chip comprising photocells, memory, clock, and an antenna. It stores in its electronic memory information identifying the sequence of an attached oligonucleotide probe. Complementary fluorescently labeled DNA targets in a biological specimen bind to the DNA probes on the surface of the microtransponders. The chip is then scanned with a laser to detect fluorescence from bound target. Scanning also activates the transponder's memory, and the unique identification number of the microtransponder is transmitted by the microchip to a receiver (this information is linked to the identity of the probe on the microchip surface). Multiplex assays can easily be performed by mixing together many microtransponders (and hence many probes) with a sample. The liquid "three-dimensional array" of DNA probes is analyzed as a linear stream of microtransponders speeding through the flow chamber of the scanner.

mRNA

This is messenger RNA and is transcribed from DNA and eventually translated into protein ("DNA makes RNA makes protein"). Microarray chips have been effective devices for analyzing complex mixtures of mRNAs.

Multiple-force-generating microchips

Structures that provide combinations of dielectrophoretic, acoustic, thermal gradient, and magnetic forces can be incorporated inside a single microchip to produce the so-called "multiple-force"-generating microchips. This type of microchip is particularly advantageous when manipulating cells and magnetic beads. Figure 33 shows an example of a bioelectronic microchip that incorporates small electromagnets and electrodes to exert both magnetic and dielectrophoretic forces on components of mixtures within the microchip.

Figure 33. A multiple-force bioelectronic microchip that incorporates small electromagnets (darker objects beneath the electrodes) and electrodes (the vertical castellated structures) to exert both magnetic and dielectrophoretic forces on components of mixtures within the microchip.

Reproduced with permission from AVIVA Biosciences Corporation, San Diego, CA.

Nanobead

Beads or particles with a diameter in the 0.1- to 100-nm range are termed nanobeads or nanoparticles. Nanobeads have a number of established applications (e.g., imaging, drug, or vaccine delivery), and reagents immobilized on very small beads have been used inside microchips for conducting binding reactions (e.g., antigen:antibody reactions).

Nanochip

This is the nanotechnological equivalent of today's microchip. It will have nm-sized components and will perform analysis in ways as yet unknown. Although nanochips are not quite ready for prime time, devices with nL volumes for biochemical analysis are available now. The array of miniaturized reaction chambers shown in Figure 34 operates using

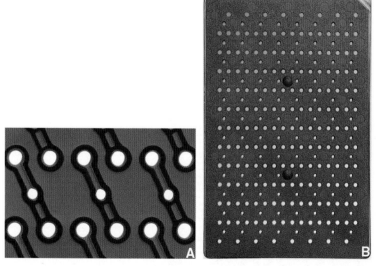

Figure 34. Microfluidic device with nL capacity (Arteas™, ACLARA BioSciences, Mountain View, CA). Individual devices (A) are arrayed in a 96-well plate format (B). The three-well structures can be seen in cross-section linked by a nL-volume channel connecting the large auxiliary reservoir well (upper) to the middle reaction well.

nL volumes (400–600 nL) or reagents. At these ultra-low volumes, evaporation is a major concern. This is obviated by a large auxiliary reservoir well that constantly replenishes the reaction well via a small channel.

Nanoelectronics

Conventional transistor technology will eventually reach its minimum size. The current view is that nanoelectronics will take electronic devices past that limit and permit miniaturization at the nanoscale. Nanoelectronic components, dominated by quantum mechanical effects (e.g., tunneling of electrons), are slowly emerging as viable replacements for their microscale counterparts. These nm-sized devices include quantum dots (artificial atoms), single-electron transistors, carbon nanotubes, and molecular wires. These incredibly small devices could be easily integrated as components of analytical microchips to provide electronic control of the analytical process, data crunching, and communication.

Nanoliter

One-billionth of a liter (10^{-9} L). A teaspoon is a 5-mL measure of liquid and this is equivalent to 5 million nL.

Nanometer

One-billionth of a meter (10^{-9} m). Engineers often use a mil (one-thousandth of an inch) as a unit of measure, and this is equivalent to 25,400 nm.

Nanoparticle [See "Nanobead"]

Nanotechnology

The science of nanotechnology is concerned with materials and systems whose structures and components exhibit novel and significantly improved physical, chemical, and biological properties, phenomena and processes because of their small nanoscale size. Structural features in the range 10^{-9} to 10^{-7} m (1–100 nm) determine important changes as compared to the behavior of isolated molecules (1 nm) or of bulk materials (http://www.house.gov/science/smalley_062299.htm).

An alternative definition extends nanotechnology to one-tenth of a nanometer: "The branch of technology that deals with dimensions and tolerances of 0.1–100 nm, or with the manipulation of individual atoms and molecules" (source: OED online at http://dictionary.oed.com).

Nanotube

Also known as a buckytube, microtubule, or fibril. These are nanometer diameter tubes (10,000 times thinner than a human hair!) of varying length, made from hexagonal arrangements of carbon atoms (graphite) attached at the edges to form one or more concentric tu-

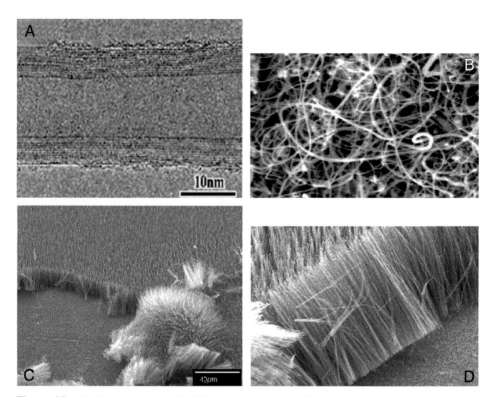

Figure 35. Carbon nanotubes: (A) TEM image of multi-wall carbon nanotubes that shows a hollow center of about 10 nm; (B) SEM image of carbon nanotubes grown by chemical vapor deposition; (C) SEM image of carbon nanotubes grown by plasma-enhanced chemical vapor deposition, which shows the very good alignment perpendicular to the substrate surface.

bular structures (Figures 35A and 35B). They can be produced in a variety of forms, including the ordered arrays on the surface of a flat substrate, as shown in Figure 35C and 35D.

Nanotubes are an important nanotechnological device and have already found application as tips for AFMs and in a variety of electronic devices (e.g., display panels). Additionally, a molecule can be attached to the tip of a nanotube to make it bio-specific. When it is moved across a surface, this tiny device senses bio-specific interactions dictated by the molecular recognition properties of the molecule attached to the nm-sized tip. This type of analysis is called scanometric analysis. For more information, visit "The Nanotube Site" at http://www.pa.msu.edu/cmp/csc/nanotube.html.

Reproduced with permission from Boston College's paper and NanoLab, Inc., Brighton, MA (web page: www.nano-lab.com) (*17,18 [Figures 35A–D]*). *Reprinted with permission from Ren et al. Synthesis of large arrays of well-aligned carbon nanotubes on glass.* Science *1998;282:1105–7).* © *1998: American Association for the Advancement of Science (Figure 35C and 35D).*

Ode

This "Ode to a microchip" is an exercise in alliteration in its purest form. All words start with the same letter, and no word is used more than once:

Microchip
Majestic monolith
Mighty multipurpose microanalyzer
Merciless micronic mover
Massively parallel microassay monitor
Mass-produced money-maker
Munificent-micromachined motivator
Minuscule microspots marshall misguided mRNA
Minute microelectrodes mandate migratory movement
Miniature manipulators move motile micro-organisms
Monumental microfluidic manifolds merge murky mixtures
Massed microfilters mesh mutilated materials
Meticulous microdispensers mete meager morsels
Meandering microchannels maneuver massive molecules
Mesoscale microchambers mingle mangled microbes
Magnanimous
Manifestly minimalistic
Misnamed, misunderstood micromachine
Masked manufacture
Microcomputer mother
Mankind's multipurpose marvel
Millennium's microetched masterpiece

Oligonucleotide microarray [See "Microarray"]

An array of oligonucleotides (usually in the size 6- to 25-mer, deposited or synthesized in situ on a surface of a substrate (a silicon chip, glass microscope slide, or a sheet of plastic). Oligonucleotide microarrays have found numerous applications including re-sequencing, gene expression analysis, mutation detection, gene copy number determination (LOH or loss of heterozygosity), genotyping, gene discovery, forensic analysis, and toxicity assessment.

Omes and omics

The suffixes of the 1990s and the new millennium. As a general rule, for every -ome (derived from the Latin for "mass") there is an -omic. For example, biome/biomics, genome/genomics, cellome/cellomics, cytome/cytomics, functome/functomics, oper-ome/operomics (any singing involved?), proteome/proteomics, etc.! I recommend that you visit http://www.genomicsglossaries.com for an informative -omes and -omics glossary, and also http://www.the-scientist.com/yr2000/apr/comm_010402html for "Ome Sweet Omics: A Genealogical Treasury of Words".

Passivation [See "PDMS"]

A surface can have properties that render it chemically or physically reactive towards other substances, and this is detrimental in certain analytical applications. It can be particularly troublesome in microchips that have high surface-to-volume ratios (e.g., surface-to-area ratio of a 8.32-μL microchip is >25 as compared to 1.5 for a conventional 50-μL PCR tube). Adverse surface chemistry effects can be eliminated by passivating the surface by applying a coating of an inert material. In the case of silicon PCR microchips, a layer of silicon dioxide has been found to passivate the silicon surface which otherwise would interfere with the PCR reaction.

Patents

Few would dispute Abraham Lincoln's contention that "The patent system added the fuel of interest to the fire of genius." In the United States the basis for the Federal patent law can be found in the United States constitution drafted in Philadelphia in 1787. Article 1, Section 8, Clause 8 states that Congress will have the power to "promote the progress of science and useful arts by securing for limited times to authors and inventors the exclusive right to their respective writings and discoveries." Subsequent Patent Acts (1790, 1793) defined what is patentable as "any new and useful art, machine, manufacture or composition of matter and any new and useful improvement on any art, machine, manufacture or composition of matter," and this view has remained largely unchanged (http://ladas.com/uspatenthistory.html). Other countries have a longer history of patenting. In Great Britain, for example, the earliest known English patent was granted by Henry VI to Flemish-born John of Utynam in 1449—it gave him a 20-year monopoly for a method of making stained glass . In those days patenting was a very bureaucratic process: it required visiting no less than seven offices and obtaining two signatures by the monarch (a tall order at the best of times!) (http://www.patent.gov.uk/patent/history/fivehundred/origins.htm).

United States Patent No. 1 was issued on 13 July 1836, to John Ruggles of Thomaston, Maine, for "locomotive steam-engine for rail and other roads." More than 150 years later, enthusiasm for patenting continues unabated, and over 6 million US patents have now been granted.

The commercial importance of microchip technology and the considerable inventive effort has produced hundreds of patent applications and patents. If you want to explore the many issued patents and patent applications, then the most accessible and easy-to-use patent office website is the United States Trademark and Patent Office (USTPO) site at http://www.uspto.gov (and its free!). Also visit the UK patent office at http://www.patent.

Figure 36. Group of silicon–glass PCR and PCR cell filtration microchips (14 mm × 17 mm). The large chamber on each microchip is connected by two small channels to exit and entry ports. The structures traversing the chambers on some of the microchips are microfilters for cell isolation.

gov.uk/, the European Patent Office at http://european-patent-office.org/, and http://www.pcug.org.au/~arhen/ for a compilation of 50 other patent office sites.

PCR microchip

Silicon heats up and cools down rapidly, and its high thermal conductivity (2.33×10^{-6} °C^{-1}), makes it an ideal material for thermal cycling applications such as the PCR. Microminiature chambers etched into silicon and capped with glass (PCR microchip) are suitable for microscale PCR (Figure 36), but oxide treatment to passivate the native silicon microchamber surface is required in order to prevent interference in the PCR.

PDMS

Poly(dimethylsiloxane) (PDMS) is a polymer used to fabricate PDMS microfluidic networks. Depending on the application, the hydrophobic PDMS surface can be modified by treatment with a polybrene–dextran sulfate bilayer (capillary electrophoresis) or a three-layer biotinylated-IgG–neutravidin–biotinylated-dextran sandwich (immunoassay). Molded PDMS microchips capped with glass are effective as single-use PCR microchips.

Personal laboratory

One expected outcome of the application of microminiaturization to analytical techniques is the emergence of the personal laboratory. This is envisaged as a small, hand-held, easy-to-use device that would permit ordinary people to measure substances in their own blood or urine, thus providing unparalleled access to biochemical data. It is not clear, in this futuristic scenario, if the advent of a personal laboratory would toll the death knell of the centralized laboratory. However, a trip to the neighborhood pharmacy or supermarket shows the telltale signs of progress towards this endpoint in the diverse range of pregnancy test devices and glucometers filling the shelves.

Pharmacogenomics

This is the use of genomic information to predict drug action. Variations in genes for drug metabolizing enzymes (e.g., CYP family of genes) influence drug action, and produce significant inter-individual variation in drug efficacy. DNA microarrays provide a convenient tool for studying variants of the many genes involved in drug metabolism.

Photolithography [See "Etching"; "Photoresist"]

This is a multi-step process in which the surface of a silicon wafer is etched according to a design transferred onto the wafer surface photographically using a mask. The steps in the process are as follows (Figure 37):

1. The wafer is first oxidized to form a thin surface layer of chemically resistant silicon dioxide.
2. A thin film of a photosensitive and etch-resistant material (photoresist) is spin-coated onto this surface. The mask is placed over the photoresist and the wafer exposed to ultraviolet light. This transfers the mask pattern into the photoresist. There are two types of photoresist. In the case of a positive resist, areas of photoresist exposed to light are solubilized, whereas for a negative resist, photoresist exposed to light is insolubilized.
3. Exposed areas of the positive photoresist are removed to expose the underlying oxide surface.
4. Exposed areas of silicon oxide are dissolved away with HF to produce the mask-defined pattern in the wafer surface.
5. The remaining photoresist is removed to reveal a silicon oxide replica of the mask pattern.
6. Exposed silicon is etched using either an isotropic etchant (e.g., $HF–HNO_3$–acetic acid mixture) or an anisotropic etchant (e.g., KOH; hydrazine–water; ethylenediamine–pyrocatechol–water).

For a structure requiring more than one mask, steps 2–6 are repeated until etching of the entire structure is completed.

7. In the final step, the oxide layer is removed to reveal the etched structure corresponding to the pattern transferred from the original mask.

Photoresist

This is a photosensitive material that is applied to the surface of a substrate, such as silicon or glass, and then exposed to radiation (e.g., ultraviolet light, X-rays) through a mask that carries the specific pattern to be etched into the substrate surface. Exposed areas of a positive resist (e.g., Novolak) become soluble and are removed by solvent treatment. In the case of a negative resist, the exposed areas are insolubilized by a photochemical cross-linking reaction, and the soluble unexposed areas are removed by solvent treatment.

Picoliter

One-trillionth (10^{-12}) of a liter. The liquid measure, the "dram" (as in a dram of single malt Scottish whisky!), is a mere 1 oz of whisky, and this is 1/5 of a gill, and a gill is 1/4 of a pint. So the one dram in your glass is 28.04 mL, which is the same as 28.04 billion pL. If you ordered a "dram" in the United States you would get less—in fact you would only have 3.69669 billion pL in your glass! Also, beware you might be served the Irish or American beverage called whiskey! (see http:/members.tripod.com/anquaich/whisky.htm).

Pins

Spotting of probes (or targets) to form a microarray is performed using a pin or groups of pins that are dipped into a solution of a probe and applied to the surface of a

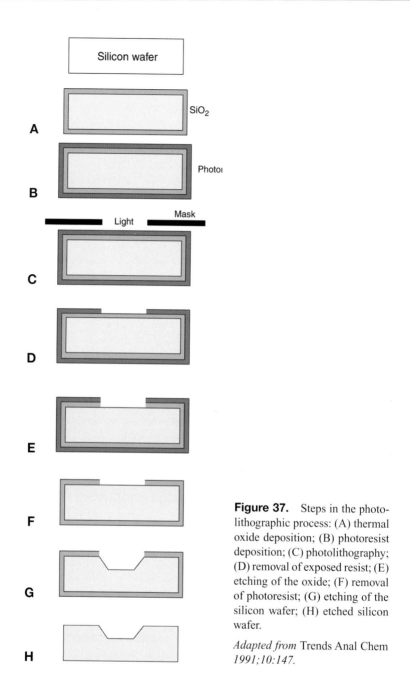

Silicon wafer

A — SiO₂ — SiO_2

B — Photoı

Light — Mask

C

D

E

F

G

H

Figure 37. Steps in the photo-lithographic process: (A) thermal oxide deposition; (B) photoresist deposition; (C) photolithography; (D) removal of exposed resist; (E) etching of the oxide; (F) removal of photoresist; (G) etching of the silicon wafer; (H) etched silicon wafer.

Adapted from Trends Anal Chem *1991;10:147.*

substrate. Different designs of pins are available, including the quill and the split-pen pin that function just like the nib on an old-fashioned ink pen. It is dipped into a solution of the substance and touched to the surface of the substrate where it deposits a small amount of the solution. More of the substance can be deposited by repeated application at the same location.

Plastic microchips

A microchip fabricated in a plastic material has a number of potential benefits. It opens up a route to the low-cost, high-volume production methods widely used to produce many types of disposable laboratory tubes and tips. It also provides access to materials with differing physical and chemical properties that may be employed advantageously in the design of microchips and microchip-based assays. Currently, plastic microchips are available in PDMS, PMMA, polycarbonate, and polyimide. They have been designed mainly for electrophoresis, but have also found use in chromatography, immunoassay, coagulation, and cell analysis (Figures 38A–38H).

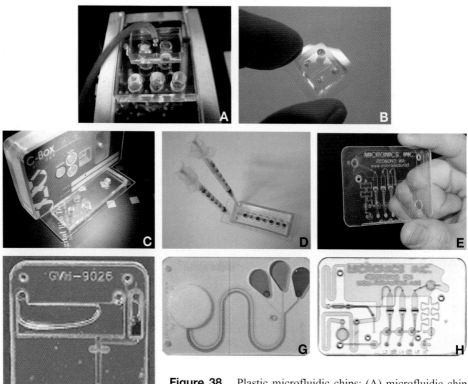

Figure 38. Plastic microfluidic chips: (A) microfluidic chip for determining prothrombin time (PT Check™); (B) erythrocyte analyzer (50–100 μm channels; 2–6 μm wide microfilters) (RBC Hemocytometer™); (C) cell analyzer incorporating a 50-μm × 100-μm microanalytical channel (C-Box™); (D) automated serial dilution device (Dilution Palette™); (E) Micronic's ORCA™ lab card; (F) Micronic's H-FILTER® lab card; (G) Micronic's T-sensor® lab card; (H) Micronic's Microcytometer™ lab card.

Reproduced with permission from Digital Biotechnology Co. and Prof. Jun Keun Chang, Seoul National University (Figures 38A–38D); MICRONICS, Inc., Redmond, WA (Figures 38E–38H).

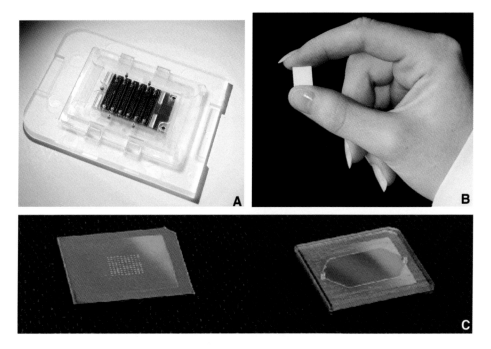

Figure 39. Protein microarrays: (A) the Zyomyx protein profiling Biochip™ contains six 15-μL sample chambers that can each measure up to 200 different proteins; (B) Randox® Laboratories biochip array (1 cm²) comprising a 10 × 10 array of test sites for panels of immunological tests (e.g., tumor markers, antibiotics, drugs of abuse); (C) HTS Biosystems protein array chip (FLEX CHIP™; 1 in.²; ~250-μm spots on 500-μm centers) and assay cartridge (right).

Reproduced with permission from Zyomyx Inc., Hayward, CA (Figure 39A); Randox® Laboratories Ltd., Crumlin, UK (Figure 39B); HTS Biosystems, Hopkinton, MA (Figure 39C).

PMMA

The polymer, PMMA, is useful for fabricating plastic chips by casting methods. It has been used extensively to make PMMA microchips for a range of microchannel-based electrophoretic separation methods (e.g., isotachophoresis).

Printing

Johann Gutenberg (1400–1468), the German inventor of movable type, could not have envisaged that the printing technology he pioneered would one day be used to make microarray devices for testing and analyzing biological fluids. All types of modern-day printing techniques have been adapted to fabricate microarrays including stamping, and ink-jet printing.

Probe

By convention, in the context of microarrays, this is the molecule immobilized on the array surface that reacts with target in the sample applied to the array (*19*).

Protein chip [See "Protein microarray"]

General term for microchips used for analysis of protein mixtures.

Protein microarray

Protein and peptide arrays are useful for high-throughput immunoassays, and assays designed to characterize protein–protein and protein–nucleic acid interactions. They can be made by deposition or by in situ synthesis. The protein microarray chip is a key tool in drug discovery and in the emerging discipline of proteomics (Figures 39A–39C).

Proteomics [See "Protein microchips"]

This is the study of all the proteins expressed by a genome. The estimated 30,000 genes of the human genome translate into between 300,000 and 1 million proteins when post-translational modifications and alternate splicing are taken into consideration. Analytical tasks of this magnitude cannot be easily contemplated using conventional techniques, but luckily, the new protein microarrays provide the parallel analyses and high-throughput capabilities needed to meet this challenge.

Quantum dots

A quantum dot is an intensely luminescent nm-sized semiconductor nanocrystal (CdSe, InAs, SiGe) used as label in DNA probe assays and immunological assays. The main advantages of quantum dots over conventional organic fluorescent dyes are increased brightness, stability against photobleaching, a broad continuous excitation spectrum, and a narrow emission spectrum. The nanocrystals can be embedded in organic matrices, such as latex beads, to produce families of beads of different colors for use as labels in multiplexed assays.

Quartz

Mono-crystalline quartz is a form of silica (SiO_2). It can be micromachined to produce many different types of microdevices, including pressure sensors and accelerometers based on detecting vibration or deflection of micromachined quartz structures (e.g., tuning forks, microbeams).

Quotations

Some quotations that seem appropriate to the area of microchips include:

"Chips with every damn thing. You breed babies and you eat chips with everything." (Arnold Wesker)

"Small is beautiful." (Proverbs)

"The best things come in small packages." (Proverbs)

"How very small the very great are!" (William Makepeace Thackeray)

"He was not merely a chip of the old block, but the old block itself." (Edmund Burke)

"For who hath despised the day of small things?" (Zechariah)

"In small proportions we just beauties see." (Ben Jonson)

Reactive ion etching

This type of etching uses fluorine-containing gases (CF_4, SF_6) in conjunction with a plasma source (radio frequency or inductively coupled plasma). It is relatively slow (a few µm/h) but can achieve structures with very high aspect ratios (etch depth/width > 40).

Refrigerator on a chip

Tiny 40-µm $\times$ 40-µm thermoelectric refrigerators can be built in precise locations directly on top of microelectronic chips to cool the underlying circuitry. They may also have many other potential applications in microchip-based analyzers, e.g., to stabilize on-chip reagents.

Reverse dot blot

An analytical strategy used in many types of microarrays in which the analytical reagent, i.e., the probe, is immobilized and reacted with a sample of the test specimen containing the target DNA (Figure 40B).

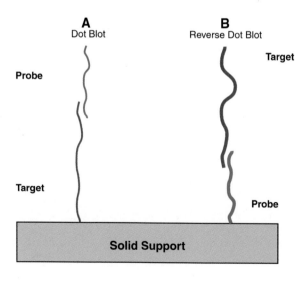

Figure 40. The dot blot and reverse dot blot analytical strategies: (A) dot blot—the target is immobilized and this reacts with probe in solution; (B) reverse dot blot—the probe is immobilized and this reacts with target in solution.

Scanner [See "Two-color analysis"]

An instrument used to measure the signal, usually fluorescence, from individual locations on a microarray. Two-color scanners are useful for scanning expression arrays and acquiring signal from two different target populations (test and reference sample). Four-color scanners are useful for array-based single nucleotide extension (SNE) assays that involve extension of hybridized probes with one of the four dideoxynucleotides, each labeled with a different fluorophore.

Self-assembled monolayer (SAM)

Highly ordered, two-dimensional arrays produced by molecular self-assembly represent a new direction in microchip fabrication. A SAM is made by exposing a metal surface, such as gold, to a long-chain alkyl thiolate, e.g., dodecanethiolate. The thiolate reacts with the metal surface according to the following equation:

$$Me(CH_2)_{11}SH + Au(0) \rightarrow Me(CH_2)_{11}S - Au(I) + \tfrac{1}{2}H_2$$

Once attached to the gold surface, van der Waals forces orient the molecules into a closely packed array. The molecules can be arranged in a pattern on the surface by microcontact printing or micropen writing. The resulting SAM can be used as an ultra-thin resist for etching planar or curved surfaces to form μm–nm-sized features. A mixture of $K_2SO_3/K_3Fe(CN)_6/K_4Fe(CN)_6$ is usually employed as an etchant for a thiolate-based SAM on a gold surface (the SAM-covered areas etch at a different rate compared to areas of the substrate that are uncovered).

Sequencing by hybridization (SBH)

A sequence of DNA can be considered an overlapping series of shorter oligonucleotides, and this notion has led to an innovative sequencing method based on hybridization to a microarray of immobilized oligonucleotides (20). Table 4 illustrates how a sequence can be reconstructed using immobilized 8-mers. A positive hybridization signal with the oligo TACCTGAC signifies the presence of the complement to this sequence in the unknown sequence (Table 4A). The array contains the four 8-mers extended by one base in the 3'-direction, and this allows interrogation of the next position in the unknown sequence. A positive result with the site on the array containing the oligo ACCTGACC re-

Table 4. Sequencing by Hybridization

Unknown sequence:	ATGGACTGGACCTGGAGGAGTCCTTCTGTC

A. Interrogation of next base

Probe on array	Hybridization result
TACCTGAC	+
ACCTGAC**A**	−
ACCTGAC**T**	−
ACCTGAC**G**	−
ACCTGAC**C**	+

B. Reconstruction of sequence from hybridization results

Identity of oligonucleotide probes on array giving a positive hybridization signal:

Oligo #22	TACCTGAC
Oligo #8	ACCTGACC
Oligo #1947	CCTGACCT
Oligo #26	CTGACCTG
Oligo #886	TGACCTGG
Oligo #654654	GACCTGGA

Reconstructed complementary sequence in bold:	**TACCTGACCTGGA**CCTCCTCAGGAAGACAG

veals that the next base in the complementary sequence is C. A computer assembles all the positive hybridization results, and from this data reconstructs the unknown sequence as shown in Table 4B. To perform this type of sequencing reaction, the array must contain all oligonucleotides of the length chosen to match the complexity of the target to be sequenced. The frequency of occurrence of a particular sequence of n bases is 4^{-n}. Thus, for the haploid human genome (3×10^9 bases), a 16-mer represents a sequence that will on an average occur only once ($3 \times 10^9 \times 4^{-16} = 1$); by increasing the sequence length to a 20-mer the average occurrence is reduced to 0.3% (*21*). So how many oligonucleotides are needed for SBH? For an oligonucleotide of length n, the number of different oligonucleotides possible is 4^n, so there are 16,384 7-mers, 65,536 8-mers, 262,144 9-mers, over 1 million 10-mers, and more than one trillion 20-mers!

In practice the task of assembling all trillion 20-mers has not been attempted; instead, smaller subsets of oligonucleotides are chosen for re-sequencing relatively short target sequences. The first clinical application of SBH was for sequencing the HIV-1 clade B protease gene (*pr* gene) in viral isolates to assess the presence of mutations important in resistance to protease inhibitor therapy. The microarray consisted of 12,224 different oligonucleotides (11- to 20-mers) in 93-μm × 93-μm–sized locations on a 1.28-cm² glass microchip (*22*).

Sequencing chip [See "SBH"]

SHOM [See "SBH"]

Abbreviation for "sequencing by hybridization with oligonucleotide matrix".

Table 5. Current List of Names and Symbols of Decimal Multiples and Sub-multiples of SI Units (see http://ts.nist.gov for more information)

Prefix	Symbol	Multiplier	
yotta-	Y	1 000 000 000 000 000 000 000 000	10^{+24}
zetta-	Z	1 000 000 000 000 000 000 000	10^{+21}
exa-	E	1 000 000 000 000 000 000	10^{+18}
peta-	P	1 000 000 000 000 000	10^{+15}
tera-	T	1 000 000 000 000	10^{+12}
giga-	G	1 000 000 000	10^{+9}
mega-	M	1 000 000	10^{+6}
kilo-	k	1 000	10^{+3}
hecto-	h	100	10^{+2}
deca-	da	10	10^{+1}
		1	
deci-	d	0.1	10^{-1}
centi-	c	0.01	10^{-2}
milli-	m	0.001	10^{-3}
micro-	μ	0.000 001	10^{-6}
nano-	n	0.000 000 001	10^{-9}
pico-	p	0.000 000 000 001	10^{-12}
femto-	f	0.000 000 000 000 001	10^{-15}
atto-	a	0.000 000 000 000 000 001	10^{-18}
zepto-	z	0.000 000 000 000 000 000 001	10^{-21}
yocto-	y	0.000 000 000 000 000 000 000 001	10^{-24}

SI

In 1960, the Système International d'Unités (SI) Conférence Générale des Poids et Mesures (CGPM) adopted a series of prefixes and symbols of prefixes to form the names and symbols of decimal multiples and sub-multiples of SI units (Table 5).

Silicon [See "Etching"]

A non-metallic element (atomic number 14) in Group IVB of the Periodic Table. Wafers (0.5- to 0.7-mm thick) are cut from a zone-refined single-crystal silicon ingot (e.g., 100-, 125-, or 150-mm diameter; impurity level $<10^{-10}$) most commonly in crystal plane orientation <100> or <111>, and are used as the raw material for fabricating many types of analytical and microelectronic chips. Silicon reacts with a variety of chemicals, such as potassium hydroxide , and this is exploited in wet chemical etching of silicon.

SNE

Single nucleotide extension (SNE) is a method for determining the identity of a base in a target DNA molecule. The target is hybridized to a probe attached to a solid surface, e.g., on a microarray. The free end of the probe is then extended by one base using dideoxynucleotides, each labeled with a different color fluorescent dye, in a PCR reaction. The base incorporated will be complementary to the base on the target adjacent to the end of the probe, and because dideoxy derivatives are used in the reaction, no further addition of bases can occur. So, the color of the fluorescence from the spot on the array will signal the identity of the complementary base on the target.

SNP

A single nucleotide polymorphism (SNP) is a single-base change at a specific position in the genome, in most cases with two alleles. They are found at a frequency of about 1 at every 1000 bases and represent the largest quantity of inter-individual genetic variation. SNP assays can conveniently be performed using oligonucleotide microarrays. Characterizing this vast number of sequence variations is a gargantuan task, but the massively parallel high-throughput analytical characteristics of oligonucleotide microarrays and bead arrays offer hope even for a task of this magnitude.

Sperm chip

Microchannels provide a convenient analytical tool for assessing the motility of sperm in a semen sample. A small sample of semen is introduced into the chamber at one end of the long and tortuous narrow channels in a sperm chip (Figures 41A and 41B). The sperm swims along the channel, and the time that they take to complete this mini-obstacle course is a measure of their motility.

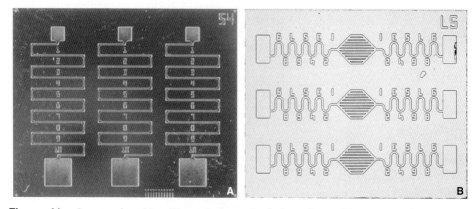

Figure 41. Sperm microchips. The etched μm-sized channels provide raceways for sperms to swim along. The winner is the fastest (most motile) sperm. A. Silicon sperm microchip (14 mm × 17 mm; shown uncapped) with three identical microchannels (80-μm wide × 20-μm deep) linked to square chambers. (B) Glass sperm microchip (14 mm × 17 mm; shown uncapped) with three identical structures. The central grooved sperm application chamber links via two identical 100-μm wide × 40-μm deep tortuous microchannels (for assays in duplicate) to square chambers at the end of each channel. Numbers etched into the chip surface provide a scale for assessing the progress of sperm along each of the microchannels when viewed through a microscope.

Spotting [See "Microarray"; "Pins"]

Term used to describe the process of depositing minute volumes of reagents onto a substrate to form an array of microspots.

Substrate

Term used to describe the material fabricated into a microchip (e.g., a piece of silicon wafer), or the material onto which a microarray is printed or synthesized (e.g., a microscope slide).

Surface chemistry [See "Passivation"]

The chemical properties of the internal surfaces of a microchip can directly affect biochemical reactions conducted within the chip. Understanding and controlling surface chemistry effects are vital for effective and efficient microchip performance, and are particularly important for microchips because of their high surface-to-volume ratios. Surface properties of silicon are modified by deposition of an oxide layer, and for PDMS, surface properties are modified depending on the analytical application, using a polybrene–dextran sulfate bilayer or a three-layer biotinylated-IgG–neutravidin–biotinylated-dextran sandwich. In microarray manufacture, glass surfaces are modified to enhance the binding of cDNA molecules using polylysine, or long-chain aliphatic molecules that bind proteins by reverse-phase interactions are employed to affix proteins to aluminum surfaces.

Systems-on-a-chip

In the MEMS world, this conveys the concept of integrating micromechanical, sensing and electronic components on a single microchip.

Target

In the context of microarrays, this term describes a molecule in a sample applied to an array. The immobilized molecules that comprise the array are termed "probes" (*19*).

Test tube on a chip

The test tube is perhaps the most readily recognized item of analytical equipment (a laboratory icon!). The phrase "test tube on a chip" is employed by some authors to convey the idea of miniaturization of chemical assays.

Thermal bonding

This is a process used for bonding pieces of glass in the manufacture of glass microchips. In a typical thermal bonding procedure, the two glass parts of the microchip are clamped together and then subjected to a heating cycle in an oven for several hours, followed by natural cooling.

Thermal gradient microchip

The optimum temperature for hybridization of a probe and target is a function of sequence (sequences rich in G + C content melt at higher temperatures). For an array of immobilized probes, this poses the problem of selecting an appropriate hybridization temperature that preserves as much specificity as possible in the hybridization of the diverse set of probes and their targets. One way of overcoming this is to independently control the temperature at each probe site using an array of up to 100 miniature heaters (500 μm × 500 μm) on an ~27-mm × ~27-mm microchip (Figure 42).

Thin film deposition

Thin films are important components of microfabricated devices. Films can be used to passivate surfaces, form etch masks, and fabricate electrodes and membranes. The two principal film-producing methods are chemical vapor deposition (CVD) and plasma-enhanced chemical vapor deposition (PECVD). The typical CVD process reacts SiH_4 with different species at low pressure and high temperature to produce surface films of silicon dioxide, silicon nitride, or silicon carbide. Thermal evaporation or sputtering is another method useful for forming thin films of chromium, gold, or other materials.

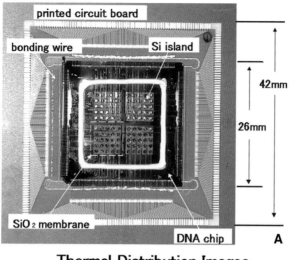

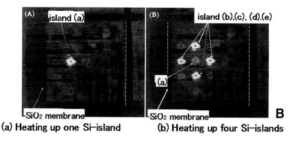

Thermal Distribution Images

(a) Heating up one Si–island   (b) Heating up four Si–islands

Figure 42. Thermal gradient chip (~27 mm × ~27 mm). DNA probes are immobilized on an array of 100 independently controlled heaters (500-μm × 500-μm islands) that control the temperature of hybridization reactions at each location on the array (A). Thermal distribution images (B) illustrating selective heating of individual islands.

Reproduced with permission from Hitachi, Ltd., Tokyo, Japan.

Tissue array

A tissue array is produced by punching out small areas of tissue from different tissue sections or different parts of a tissue section and assembling them as an array (Figure 43). This format facilitates simultaneous immunohistochemical analysis of the individual tissue sections for the presence of different antigens or in situ hybridization analysis for specific nucleic acid sequences.

"Top-down" manufacture

A manufacturing process that starts with a large block of material that is machined or otherwise processed into small microchips or nanochips. Production of a silicon–glass microchip, starting from a sheet of glass and a silicon wafer, is a good example of top-down

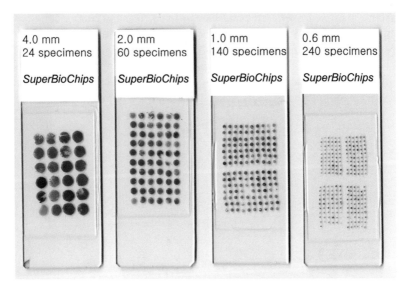

Figure 43. Tissue arrays assembled from 0.6- to 4-mm diameter disks punched from tissue sections (24–240 tissues/array).

Reproduced with permission from SuperBioChips, Seoul, Korea.

fabrication. Both components originate from larger precursors. The small glass component is cut from the larger glass sheet, and the small silicon chip component is diced from the much larger etched silicon wafer, which in turn was sliced from a solid silicon ingot.

Trademark

Many chip-based names have been protected by United States trademarks. A search of the USPTO database (http://tess.uspto.gov) for trademarks including the word "chip" yields over 2000 hits! Some examples of "live" trademark-protected names include: ASSAYCHIP (PriTest, Inc.); AVIChip (AVIVA Biosciences Corporation); CHEMCHIP (Aclara Biosciences, Inc.); CHIPLAB (Fin Scan Oy, Corp.); DIAGCHIP (Novel Science International GmbH); DNACHIP (CuraGen Corporation); FEVERCHIP (Industrial Technology Research Institute Corporation); GENECHIP (Affymetrix, Inc.); IMMUNOCHIP (National Cancer Institute); LABCHIP (Caliper Technologies Corporation); LAB-ON-A-CHIP (ABB Instrumentation Limited); LIVING CHIP (BioTrove, Inc.); MICROCHIP (Kaden Arnone, Inc.); MAGIChip (University of Chicago, Chicago, IL); MEMBRANECHIP (Proteomics Systems, Inc.); MICROCHIPS (MicroCHIPS, Inc.); NANOCHIP (Nanogen, Inc.); PICOCHIP (Picogram, Inc.); PROTEINCHIP (Ciphergen, Inc.); SNPCHIP (SNPChip, LLC); SUPERCHIP (Erie Scientific Company); TESTCHIP (TestChip Technologies, Inc.); TOXICHIP (Zyomyx, Inc.); TOXICOLOGYCHIP (Zyomyx, Inc.).

Two-color analysis [See "Expression monitoring"]

Simultaneous comparative analysis of two different samples can be achieved using two different labels that fluoresce at different wavelengths. Two-color analysis has been particularly effective in microarray-based expression analysis. Test and control cell-derived populations of mRNA or cDNA are labeled with different fluorophores, e.g., Cy3 and Cy5, or fluorescein and phycoerythrin, and applied to an oligonucleotide microarray. Fluorescence scanning reveals binding of the two different populations of fluorophore-labeled molecules at sites on the array. The intensity of the fluorescence signals at each site is proportional to the relative expression of the particular gene in the two cell populations.

Universal microchip [See "Zip-code microarrays"]

A universal DNA microchip has been proposed for both quantitating gene expression levels and for finding logical formulae for the pattern of gene expression. Universal DNA microarrays have also been designed based on artificial 24-base oligonucleotides (the so-called "zip-code microarray"). These are carefully selected so that stringent hybridizations with complementary zip-code probes can be performed at a single temperature for all probes in the microarray.

VLSIPS™ [See "Light-directed synthesis"]

Abbreviation for "very large-scale immobilized polymer synthesis."

Wafer [See "Silicon"]

A silicon wafer is a mm-thin circular slice cut from a zone-refined, single-crystal silicon ingot.

Web site

In our new virtual world the web has become an important and globally accessible source of information. Leming Shi maintains a premier website for information on all types of microchips and microarrays at http://www.gene-chips.com. Another useful site is http://www.lab-on-a-chip.com, which contains news, new products, publications, and links to items related to the miniaturized laboratory (e.g., the UK-based Lab on a Chip Consortium).

X-ray lithography

A type of lithography that uses high-energy electromagnetic radiation to define a pattern to be etched onto the surface of a silicon wafer. The short wavelength of the X-rays (4–14 Å) allows very small feature sizes (0.1 μm) to be attained.

Yeast array

An array of the yeast genes, represented by cDNA or oligonucleotides, immobilized on a chip.

Zip-code microarray

The original ZIP (zone improvement plan) code was introduced by the United States Post Office Department in 1963. The idea of a location identified by a specific code has found a parallel in microarrays in the form of the zip-code microarray. This is an array formed from a series of artificial 24-base oligonucleotides that serves as a universal platform for molecular recognition. The sequences are carefully selected so that stringent hybridizations with complementary zip-code probes can be performed at a single temperature for all probes in the microarray.

Assay protocols that involve formation of a product that incorporates the complementary zip-code probes facilitates a common hybridization reaction to the array and subsequent detection of signal at locations linked to the complementary zip-code. For example, a ligase chain reaction (LCR) can be quantified if one of the probes is labeled with a complementary zip-code oligonucleotide and the other probe with a fluorophore. Ligation produces a labeled product with a 5'-complementary zip-code oligonucleotide. This product will hybridize to the zip-code microarray at a predetermined site, thus identifying the specific complementary zip-code oligonucleotide. The zip-code can now be related back to the presence of a specific target that served as a template for the ligation of the pair of probes. The whole process can be thought of as follows. Each array location is a mailbox with a unique zip-code (the immobilized oligonucleotide). The complementary oligonucleotide is a letter that is addressed on the outside with a particular zip-code. When this probe is mailed to the array, it is delivered (hybridized) to the address, as determined by the zip-code!

Zone refining

Technique used to purify silicon ingots in order to make semiconductor grade silicon. The zone-refining process exploits preferential partitioning of impurities between the solid and a zone of melted silicon. The resulting purity is better than one part in 10^9.

References

1. Morrison P, Morrison P, and the Office of C & R Eames. Powers of ten. New York: Scientific American Library, 1982:150 pp.
2. Tucker AJ. Biochips: can molecules compute? High Tech 1984;79:36–47.
3. Gibbs WW. Patently inefficient. Sci Am 2001 (February);284:30.
4. Farkas DH. DNA simplified II. Washington: AACC Press, 1999:117 pp.
5. Barinaga M. Will "DNA chip" speed genome initiative? Science 1991;253:1489.
6. Drexler KE. Engines of creation. The coming era of nanotechnology. Garden City: Anchor Press/Doubleday, 1986:298 pp.
7. Feynman RP. There's plenty of room at the bottom. In: Proceedings of the talk at the annual meeting of the American Physical Society, California Institute of Technology, December 29, 1959.
8. Terry SC. A gas chromatograph system fabricated on a silicon wafer using integrated circuit technology. PhD thesis, Stanford University, 1975.
9. Terry SC, Jerman JH, Angell JB. A gas chromatographic air analyzer fabricated on a silicon wafer. IEEE Trans Electron Devices 1979:ED-26:1880–6.
10. The Celera Genomics Sequencing Team. The sequence of the human genome. Science 2001;291:1304–51.
11. The International Human Genome Mapping Consortium. A physical map of the human genome. Nature 2001;409:934–51.
12. Sharpe AN, Michaud GL. Hydrophobic-grid membrane filters: a new approach to microbiological assays. Appl Microbiol 1974;28:223–5.
13. Steber, GR. Audio frequency DSP laboratory on a chip–TMS32010. Proc IECON '87: Int. Conf Industr Electronics, Control Instrument (Cat. No. CH2484-4) 1987;2:1047–51.
14. Moser I, Jobst G, Aschauer E, Svasek P, Varahram M, Urban G, et al. Miniaturized thin film glutamate and glutamine biosensors. Biosens Bioactuators 1995;10:527–32.
15. Manz AN, Graber N, Widmer HM. Miniaturized total chemical analysis systems: a novel concept for chemical sensing. Sens Actuators B 1990;1:244–8.
16. Ekins R, Chu F, Biggart E. Development of microspot multi-analyte ratiometric immunoassay using dual fluorescent-labelled antibodies. Anal Chim Acta 1989;227:78–96.
17. Li WZ, Wen JG, Tu Y, Ren ZF. Effect of gas pressure on the growth and structure of carbon nanotubes by chemical vapor deposition. Appl Phys A 2001;73:259–64.
18. Ren ZF, Huang ZP, Xu JW, Wang JH, Bush P, Siegal MP, Provencio PN. Synthesis of large arrays of well-aligned carbon nanotubes on glass. Science 1998;282:1105–7.
19. Phimster B. Going global. Nat Gen 1999; 21[Suppl]:1.
20. Bains W, Smith GC. A novel method for nucleic acid sequence determination. J Theor Biol 1988; 135:303–7.
21. Cantor C, Smith CL. Genomics. New York: Wiley, 1999:596 pp.
22. Kozal MJ, Shah N, Shen N, Yang R, Fucini R, et al. Extensive polymorphisms observed in HIV-1 clade protease gene using high-density oligonucleotide arrays. Nat Med 1996;2:753–9.

Suggested Reading

Amoto I. Small things considered: scientists craft machines that seem impossibly tiny. Sci News 1989;136:8–10.

Angell JB, Terry SC, Barth PW. Silicon micromechanical devices. Sci Am 1983;248:44–55.

Cheng J, Kricka LJ, eds. Biochip technology. Philadelphia, PA: Harwood Academic, 2001:372 pp.

Crandall BC, Lewis J. Nanotechnology—research and perspectives. Cambridge, MA: MIT Press, 1992:381 pp.

Drexler KE. Nanosystems. New York: Wiley, 1992:556 pp.

Jordan B, ed. DNA microarrays: gene expression applications. Berlin: Springer, 2001:140 pp.

Kricka LJ. Miniaturization of analytical systems. Clin Chem 1998;44:2008–14.

Kricka LJ. Revolution on a square centimeter. Nat Biotechnol 1998;16:513–4.

Kricka LJ, Wilding P. Micromechanics and nanotechnology. In: Kost GJ, ed. Clinical automation, robotics, and optimization. New York: Wiley, 1996:pp. 45–77.

Kricka LJ, Fortina P. Microarray technology and applications: an all-language literature survey, including books, and patents. Clin Chem 2002;47:1479–82.

Kricka LJ, Fortina P. Nanotechnology and applications: an all-language literature survey, including books, and patents. Clin Chem, 2002;48:662–664.

Nanotech—the science of small gets down to business. Sci Am 2001;285:32–91[Special nanotechnology issue].

Petersen KE, McMillan WA, Kovacs GTA, Northrup MA, Christel LA, Pourahmadi F. Toward next generation clinical diagnostic instruments: scaling and new processing paradigms. J Biomed Microdev 1998;1:71–9.

Petersen KJ, Gee D, Pourahmadi F, Craddock R, Brown J, Christel L. Surface micromachined structures fabricated with silicon fusion bonding. Transducers'91. Dig Tech Papers 1991:397–9.

Rampal JB, ed. DNA arrays. Totowa: Humana Press, 2001:264 pp.

Ramsey JM, van den Berg, A, eds. Micro-total analysis systems 2001. Dordrecht: Kluwer Academic Publishers, 2001:689 pp.

Schena M, ed. Microarray biochip technology. Natick: Eaton Publishing, 2000:297 pp.

Schena M, ed. DNA microarrays. Oxford: Oxford University Press, 1999:210 pp.

Shoji S, Esashi M. Bonding and assembling methods for realizing a μTAS. In: van den Berg A, Bergveld P, eds. Micro-total analytical systems. Dordrecht: Kluwer Academic Publishers, 1995:165–79.

Southern EM. DNA chips: analysing sequence by hybridization to oligonucleotides on a large scale. Trends Genet 1996;12:110–5.

Stix G. Micron machinations. Sci Am 1992(November);267:107–17.

Sze SM, ed. VLSI technology, 2nd ed. Boston, MA: McGraw-Hill, 1988:676 pp.

Index

C-Box™, 65
cDNA, 3, 10, 19, 23, 46, 78, 83
   array, 10
CdSe, 68
cell filtration-PCR microchip, 52
cells, 12, 17, 39
   isolation, 12
CF_4, 69
channel-forming integral protein, 12
channels, 5, 73
CHEMCHIP, 77
chemical analysis-on-a-chip, 33
chemical vapor deposition, 75
chemiluminescence, 12
chip, 12, 15. *See also* microchip.
chipeener, 12
CHIPLAB, 77
chiplet, 12
chipmunk, 12
chipper, 12
chippies, 15
chippiness, 15
chippy, 15
chrome-on-glass masks, 44
Ciphergen, Inc., 77
cliches, 15–16
CO_2 laser, 39
Colossus, 16, 50
comb-type filter, 13
computers, 48
confocal laser scanning microscopy, 16
controversy, 16
convention, 66
copyright, 34
Coventor Inc., 52
CoventorWare™ ANALYZER™, 52
Critters on a Chip, 16
CuraGen Corp., 77
CVD, 75
cyanine dye, 16, 25
   Cy3, 16, 78
   Cy5, 16, 78
CYP, 62

databases, 17
dbEST, 25
dead volumes, 10
definition
   nanotechnology, 57
dendrimer, 8, 17
   structure, 9

DENSO Corp., 45
deoxyoligonucleotide, 30
design cycle, 53
DIAGCHIP, 77
Dickens, Charles, 27
dideoxynucleotides, 73
dielectrophoresis, 17
   forces, 55
diffusion, 44
Digital Biotechnology Co., 65
digital light processing, 18–19
Dilution Palette™, 65
DLP™ chip, 18, 44
DNA, 55, 69, 72
   array, 19
   chip, 19
   microchip, 79
   probe, 22, 76
   probe assays, 17, 68
   sizing, 11
   synthesis, 32
DNACHIP, 77
dodecanethiolate, 70
dogma, 25
dot blot, 69
dram, 63
Drexler, K. Eric, 19
drop, 43
drop-on-demand ink-jet, 31
Drosophila, 29
drug discovery, 20, 30, 46
dry etch process, 23

E. coli, 29
e-beam, 23
eggs, 36
electrochemical sensors, 36
electrode, 5, 12, 34, 75
   array, 22
   design, 15
electroendosmosis, 21
forces, 21
   -induced flow, 21
electrokinesis, 21
electronic
   addressing, 21–22
   hybridization, 22
   stringency, 22–23
Electronic Numerical Integrator and
     Computer, 16, 23, 49, 50
electrospray ionization, 44